Advances in High-Entropy Alloys - Materials Research, Exotic Properties and Applications

Edited by Jiro Kitagawa

Published in London, United Kingdom

IntechOpen

Supporting open minds since 2005

Advances in High-Entropy Alloys - Materials Research, Exotic Properties and Applications
http://dx.doi.org/10.5772/intechopen.93454
Edited by Jiro Kitagawa

Contributors
Yoshikazu Mizuguchi, Aichi Yamashita, Shailesh K Singh, Jiro Kitagawa, Naoki Ishizu, Shusuke Hamamoto, Yong A Zhang, Yaqi Wu, Modupeola Dada, Patricia Popoola, Ntombizodwa Mathe, Samson Adeosun, Sisa Pityana, Olufemi Aramide, Nicholus Malatji, Thabo Lengopeng, Afolabi Ayodeji, Fawzy H. Samuel, Mohamed F. Ibrahim, Hany R. Ammar, Agnes Marie Samuel, Mahmoud S. Soliman, Victor Songmene, Mohamed Gamal Mahmoud, Yasser Zedan, Herebert W. Doty

First published in London, United Kingdom, 2021 by IntechOpen
IntechOpen is the global imprint of INTECHOPEN LIMITED, registered in England and Wales, registration number: 11086078, 5 Princes Gate Court, London, SW7 2QJ, United Kingdom
Printed in Croatia

British Library Cataloguing-in-Publication Data
A catalogue record for this book is available from the British Library

Additional hard and PDF copies can be obtained from orders@intechopen.com

Advances in High-Entropy Alloys - Materials Research, Exotic Properties and Applications
Edited by Jiro Kitagawa
p. cm.
Print ISBN 978-1-83881-078-8
Online ISBN 978-1-83881-079-5
eBook (PDF) ISBN 978-1-83881-080-1

For EU product safety concerns:
IN TECH d.o.o., Prolaz Marije Krucifikse Kozulić 3, 51000 Rijeka, Croatia, info@intechopen.com or visit our website at intechopen.com.

Meet the editor

Jiro Kitagawa received a BS in Physics from Kyoto University, Japan, in 1993, and an MS and Ph.D. in Physics from the University of Tokyo, Japan, in 1995 and 1998, respectively. At the University of Tokyo, Dr. Kitagawa carried out materials research on rare-earth intermetallic compounds. In 2002 he joined Hiroshima University, Japan, as a research associate studying terahertz technologies. From 2012 to 2014 he was an assistant professor at Fukuoka Institute of Technology, Japan. He became a full professor there in 2015. His research interests include materials research on magnetic and superconducting materials. His current projects are on magnetism and superconductivity in high-entropy alloys.

Contents

Preface

High-entropy alloys (HEAs) are a new class of materials defined by crystals in which more than five elements, each with an atomic fraction between 5% and 35%, randomly occupy one crystallographic site. The concept of HEAs initially developed in single-site crystal structures such as face-centered cubic (fcc), body-centered cubic (bcc), and hexagonal-closed packing (hcp). However, the concept is now adopted in many multi-site alloys (multi-site HEAs). Due to the severe lattice distortion effect, many fcc or bcc HEAs show superior mechanical properties. The superior mechanical properties result from the cocktail effect, which means an enhancement of property beyond the simple mixture of constituent elements. The cocktail effect is also observed in a multi-site HEA; for example, outstanding thermal stability or enhancement of magnetic frustration in a high-entropy alloyed oxide. Due to the massive elemental combination of the HEA system, there are unlimited possibilities of finding new phenomena in the materials research on HEAs. The synthesis of a new single-phase or multiphase bulk sample is crucial work. In addition, fabrications of thin-film and nanocrystalline samples of well-known HEAs are important works. It is widely accepted that first-principle calculations, machine learning, and calculation of phase diagram (CALPHAD) are powerful methods for screening new compounds.

This book covers some very interesting topics concerning the mechanical, physical, and chemical properties of new HEAs, including high strength, high ductility, good thermal stability, superconductivity, exotic magnetism, and so on. It also examines potential applications of HEAs, such as coating against corrosion, biomaterials, catalysts, shape memory alloys, magnetic refrigeration materials, and more. This book provides the reader with a comprehensive overview of the frontier of materials research and the exotic properties (mechanical, physical, chemical, etc.) and exciting applications of HEAs.

The book consists of three sections. Section 1 focuses on HEA superconductors. Chapter 1 summarizes the frontier studies of multi-site HEA superconductors. The chapter focuses on HEA-type compounds with the NaCl-type, the CuAl2-type, high-Tc cuprates, and BiS2-based layered structures. In the three-dimensional structures with the NaCl-type and the CuAl2-type, the improvement of superconducting properties by introducing the HEA state is not clearly observed, whereas some interesting properties are found. However, high-Tc cuprates and BiS2-based layered HEA superconductors, characterized by two-dimensional crystal structures, exhibit improved superconducting properties due to high-entropy effects. The HEA effects depend on structural dimensionality. Chapter 2 describes the materials research on fcc and Mn5Si3-type HEA superconductors.

Section 2 is devoted to HEA composites. Chapter 3 examines an Al-B4C metal matrix composite (MMC), which is useful for shielding material for nuclear reactors. In this chapter, the authors introduce the microstructures of Al-B4C MMC samples. Furthermore, they examine the mechanical properties of samples and present the results from the viewpoint of microstructures. Chapter 4 reviews papers concerning the effect of mischmetal, La or Ce, and La+Ce additions on Al-Si cast alloys. The review discusses the microstructures and mechanical properties of multicomponent systems. In this chapter, the authors introduce the microstructures of many Al-Si multicomponent alloys. The mechanical properties of samples are well correlated with the results of microstructures. Chapter 5 summarizes the frontier studies of high-entropy superalloys (HESAs) and laser surface modification used to protect them. HESAs are potential alternatives to nickel superalloys and are good candidates for gas turbine applications. After presenting the advances of nickel superalloys and HESAs, the authors discuss the protection of superalloys, which is necessary for gas turbine applications. The authors focus on laser surface modifications, including laser surface melting, laser transformation hardening, laser surface alloying, and laser glazing. It is stressed that HESAs exhibit good oxidation resistance and high yield strength compared to traditional nickel superalloys.

Section 3 discusses the current state of HEA design. Chapter 6 surveys the material design of HEAs and explains the thermodynamic and electronic structure parameters. The authors review calculation and simulation methods such as density functional theory (DFT), CALPHAD, and machine learning. Finally, the authors discuss high-throughput screening methods for HEAs. Chapter 7 reports on the design of HEAs using Artificial Intelligence (AI). In materials research, it is expected that AI is quicker than other computational methods in the case of no physical model. The authors explain machine learning in the design of HEAs as well as discuss the methodology for implementing AI in HEAs.

I hope that many readers will be interested in this book and enter into the research of HEAs.

Jiro Kitagawa
Fukuoka Institute of Technology,
Fukuoka, Japan

Section 1

High-Entropy Alloy Superconductors

Chapter 1

Superconductivity in HEA-Type Compounds

Yoshikazu Mizuguchi and Aichi Yamashita

Abstract

Since the discovery of superconductivity in a high-entropy alloy (HEA) Ti-Zr-Nb-Hf-Ta in 2014, the community of superconductor science has explored new HEA superconductors to find the merit of the HEA states on superconducting properties. Since 2018, we have developed “HEA-type” compounds as superconductors or thermoelectric materials. As well known, compounds like intermetallic compounds or layered compounds are composed of multi crystallographic sites. In a HEA-type compounds, one or more sites are alloyed and total mixing entropy satisfies with the criterion of HEA. Herein, we summarize the synthesis methods, the crystal structural variation and superconducting properties of the HEA-type compounds, which include NaCl-type metal tellurides, $CuAl_2$-type transition metal zirconides, high-T_c cuprates, and BiS_2-based layered superconductors. The effects of the introduction of a HEA site in various kinds of complicated compounds are discussed from the structural-dimensionality viewpoint.

Keywords: superconductor, layered compounds, material design, high entropy alloy

1. Introduction

1.1 Superconducting materials

Superconductivity is a quantum phenomenon, which is characterized by zero-resistivity states in electrical resistivity and exclusion of magnetic flux from a superconductor [1, 2]. Superconductivity has provided many exotic research topics not only in the field of science but also in application of superconductors. The zero-resistivity states can achieve large-scale electricity transport with ultra-low energy loss, very high magnetic fields, which has been used in various devices like a magnetic resonance imaging (MRI) and a superconducting Maglev train. Although superconductor devises look perfect, the use of superconductors are regulated by temperature in reality because superconducting states are observed only at temperatures below a superconducting transition temperature (T_c), which is a parameter unique to the superconductor. To use superconducting devices, the system must be cooled down to low temperatures lower than the T_c of its superconducting components. Therefore, discovery of high-T_c superconductors has been desired.

In 1986, superconductivity with a high T_c in a Cu oxide $(La,Ba)_2CuO_4$ was discovered [3, 4]. Soon after the discovery, T_c of the Cu-oxide superconductor (cuprate) family reached 90 K for $RE Ba_2Cu_3O_{7-d}$ (*RE*: rare earth) [5], which is higher than liquid nitrogen temperature, and finally reached 135 K in a Hg-Ba-Ca-Gu-O system [6]. After the discovery of the cuprate family, many layered compounds have been searched for high-T_c superconductivity. In 2001, superconductivity in MgB_2 with a T_c of 39 K was reported [7]. Furthermore, in 2008, FeAs-based layered superconductors REFeAsO$_{1-x}$F$_x$ with a T_c exceeding 50 K were discovered [8, 9]. Particularly, in the cuprates and FeAs-based families, unconventional (non-phonon-mediated) mechanisms of superconductivity has been proposed to explain their high T_c [10].

A surprising discovery of superconductivity at very high T_c of 203 K in H_3S was reported in 2015 [11]. The phenomenon could be achieved at extremely high pressures above 150 GPa, the high T_c and possible conventional (phonon-mediated) mechanisms have recently been attracting many researchers in the field of condensed matter physics. Furthermore, higher T_cs have been reported in related hydrides, LaH_{10} (T_c > 250 K at ~200 GPa) [12, 13] and a carbonaceous sulfur hydride system (T_c = 287.7 K at 267 GPa) [14]. To use high-T_c superconducting states in hydrides, discovery of new superconducting hydrides which become superconductive at ambient pressure. The possible strategy to realize that is the utilization of chemical pressure effects, which are applied via chemical substitutions and sometimes work like external pressure effects. Therefore, further investigations on chemical pressure effects in novel superconductors are needed, and high-entropy alloying of compounds would be one of the routes to chemically modify the crystal structure of superconductors.

1.2 Superconductivity in high-entropy alloys

Although layered compounds had been the central topic in searching high-T_c and/or unconventional superconductors over the last three decades, recent works on new superconductors have focused on various kinds of materials, which includes complicated compounds and pure metals as well. Among them, high-entropy-alloy (HEA) superconductors have been a developing field of study [15].

HEA is an alloy possessing high configurational mixing entropy (ΔS_{mix}), which is achieved by making the alloy with more than five constituent elements with an occupancy ranging 5 to 35 at% for each element [16, 17]. Typically, ΔS_{mix} of HEA is calculated by $\Delta S_{mix} = -R\ \Sigma_i\ c_i \ln c_i$, where c_i and R are the compositional ratio and the gas constant [17], and reaches 1.5R. Due to high ΔS_{mix}, HEAs exhibit stability or high performance in high temperature and/or extreme conditions [17]. Therefore, HEAs have been extensively studied in the fields of materials science and engineering.

In 2014, Koželj et al. reported superconductivity with T_c = 7.3 K in a HEA $Ta_{0.34}Nb_{0.33}Hf_{0.08}Zr_{0.14}Ti_{0.11}$ [18]. The HEA superconductor has a bcc structure with a space group of Im-3m. In **Figure 1**, we compare the crystal structures of (a) a pure Nb metal (T_c = 9.2 K), (b) a NbTi alloy (T_c ~ 10 K), which is the mostly-used practical superconductor, and (c) HEA $Ta_{0.34}Nb_{0.33}Hf_{0.08}Zr_{0.14}Ti_{0.11}$. All those materials show superconductivity and the crystal structure type is the same. The difference between them is the mixing entropy ΔS_{mix} and T_c. Although the T_c of HEA is lower than that of the other two, it was surprising for researchers that such a disordered alloy exhibits superconductivity with bulk nature. After the discovery of Ref. [18], various HEA superconductors have been developed; material information [18–26] is listed in **Table 1**.

As shown in **Figure 2**, a superconducting transition was observed in $Ta_{0.34}Nb_{0.33}Hf_{0.08}Zr_{0.14}Ti_{0.11}$ [18]. The temperature dependence of electrical resistivity (**Figure 2(a)**) shows metallic behavior but exhibits a relatively small residual resistivity ratio (RRR) at low temperatures. This would be due to the presence of disorder, and a similar trend has been observed in various HEA-type superconductors. Bulk superconductivity could be confirmed through specific heat measurement as shown in **Figure 2(b)**. From specific heat data, it has been found that most HEA superconductors exhibit conventional (phonon-mediated) pairing states.

T_c of HEA superconductors show correlation with valence electron count (VEC) [15, 21, 27]. The type-A HEAs with lower VEC exhibit a dependence of T_c on VEC

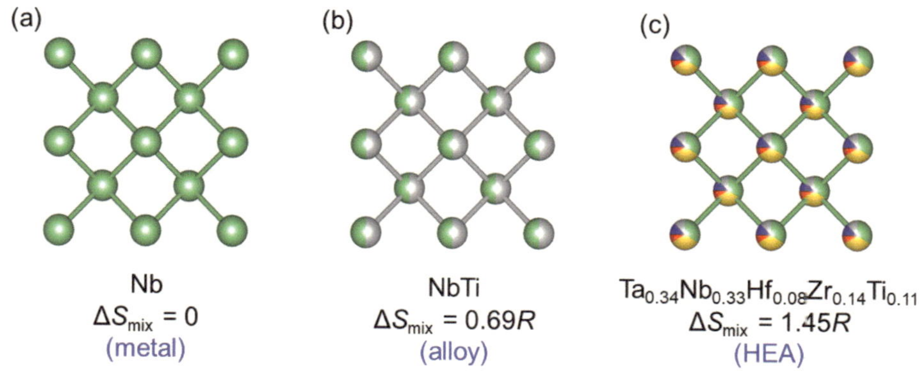

Figure 1.
Schematic image of crystal structure of (a) Nb (pure metal), (b) NbTi (alloy), and (c) $Ta_{0.34}Nb_{0.33}Hf_{0.08}Zr_{0.14}Ti_{0.11}$ (HEA).

similar to that for crystalline metals, alloys, and amorphous materials, and the T_c of HEAs are intermediate between crystalline materials and amorphous. In contrast, the T_c of the type-C HEAs with middle VEC shows opposite behavior to that for other forms. For the type-B HEAs, the trend of T_c on VEC seems resembling that for other forms, but their T_cs are clearly lower than that for crystalline and amorphous materials. As mentioned in Ref. [15], there would be a clear effect of crystallinity on T_c at the same VEC range, but the origin of the different trends between type-A, type-B, and type-C regimes have not been clarified from physical viewpoint (**Figure 3**).

Another notable character of HEA is a robustness of superconductivity to extremely high pressure. As reported in Ref. [20], the T_c of Ta-Nb-Hf-Zr-Ti slightly increases by external pressure effect and the 10 K-class T_c maintains under extreme pressures like 200 GPa. However, the robustness of superconductivity to extremely high pressure was reported for simpler NbTi with a clear increase in T_c to 19.1 K at 261 GPa [28]. Therefore, the issue if HEA can improve the stability of superconductor under extremely high pressures has not been clarified.

1.3 Concept of HEA-type compounds

As described in subsection 1–2, superconductivity in HEAs has been discovered and been regarded as a new research field of superconductivity. However, the merit of HEA states for superconductors has not been fully understood. Therefore, development of new types of HEA superconductors is needed. The hint to expand the material variation of HEA superconductors was proposed in Ref. [29], in which a HEA superconductor with a CsCl-type structure was reported. Since the CsCl structure contains two independent crystallographic sites, we have flexibility of elemental solution at the two sites. When calculating total ΔS_{mix} of $(ScZrNbTa)_{0.65}(RhPd)_{0.35}$ by taking the sum of ΔS_{mix}s for site-1 and site-2, it appears to reach very high ΔS_{mix} of 1.79R. A similar site separation has been observed in $(Nb_{0.11}Re_{0.56})(HfZrTi)_{0.33}$ [30]. Motivated by those studies on HEAs with site separation, we have tried to synthesize various "*HEA-type compounds*", which contain NaCl-type metal chalcogenides [31–33], $CuAl_2$-type tetragonal $TrZr_2$ (Tr: Fe, Co, Ni, Cu, Rh, Ir) [34, 35], high-T_c RE123 cuprates [36], and BiS_2-based layered superconductors [37, 38]. The concept of HEA-type compounds it that we achieve a high ΔS_{mix} by site-selective alloying. As shown in **Figure 4**, HEA-type compounds have a HEA-type site, in which five or more elements are solving and a normal site, which is not in the HEA state. The list of superconducting HEA-type compounds is shown in **Table 2**. By studying HEA effects to crystal structure and

Composition	$\Delta S_{mix}/R$	T_c (K)	Structure	Ref.
$Ta_{0.34}Nb_{0.33}Hf_{0.08}Zr_{0.14}Ti_{0.11}$	1.45	7.3	bcc	[18, 19]
$(TaNb)_{0.67}(HfZrTi)_{0.33}$	1.46	7.7	bcc	[20]
$(TaNb)_{0.7}(ZrHfTi)_{0.3}$	1.43	8.0	bcc	[21]
$(TaNb)_{0.7}(ZrHfTi)_{0.33}$	1.24	7.8	bcc	
$(TaNb)_{0.7}(ZrHfTi)_{0.4}$	1.31	7.6	bcc	
$(TaNb)_{0.7}(ZrHfTi)_{0.5}$	1.39	6.5	bcc	
$(TaNb)_{0.7}(ZrHfTi)_{0.84}$	1.60	4.5	bcc	
$(TaNb)_{0.67}(Hf)_{0.33}$	1.10	7.3	bcc	
$(TaNb)_{0.67}(HfZr)_{0.33}$	1.33	6.6	bcc	
$Nb_{0.67}(HfZrTi)_{0.33}$	1.00	9.2	bcc	
$(NbV)_{0.67}(HfZrTi)_{0.33}$	1.46	7.2	bcc	
$(TaV)_{0.67}(HfZrTi)_{0.33}$	1.46	4.0	bcc	
$(TaNb)_{0.67}(HfZrTi)_{0.33}$	1.46	7.3	bcc	
$(TaNbV)_{0.67}(HfZrTi)_{0.33}$	1.73	4.3	bcc	
$Hf_{0.21}Nb_{0.25}Ti_{0.15}V_{0.15}Zr_{0.24}$	1.59	5.3	bcc	[22]
$Ta_{0.35}Nb_{0.35}Zr_{0.15}Ti_{0.15}$	1.30	8.0	bcc	[23]
$(ZrNb)_{0.2}(MoReRu)_{0.8}$	1.52	4.2	bcc	[24]
$(ZrNb)_{0.1}(MoReRu)_{0.9}$	1.38	5.3	bcc	
$(HfTaWIr)_{0.6}Re_{0.4}$	1.50	1.9	bcc + hcp	
$(HfTaWIr)_{0.5}Re_{0.5}$	1.39	2.7	bcc + hcp	
$(HfTaWIr)_{0.4}Re_{0.6}$	1.23	4.0	bcc	
$(HfTaWIr)_{0.3}Re_{0.7}$	1.03	4.5	bcc	
$(HfTaWIr)_{0.2}Re_{0.8}$	0.78	5.7	bcc	
$(HfTaWPt)_{0.5}Re_{0.5}$	1.39	2.2	bcc + hcp	
$(HfTaWPt)_{0.4}Re_{0.6}$	1.23	4.4	bcc	
$(HfTaWPt)_{0.3}Re_{0.7}$	1.03	5.7	bcc	
$(HfTaWPt)_{0.25}Re_{0.75}$	0.91	6.1	bcc	
$Nb_{26.1}Ta_{25.1}Ti_{23.4}Zr_{0.254}$	1.39	8.3	bcc	[25]
$Nb_{0.198}Ta_{0.189}Ti_{0.208}Zr_{0.187}Hf_{0.218}$	1.61	7.1	bcc	
$Nb_{0.163}Ta_{0.157}Ti_{0.169}Zr_{0.171}Hf_{0.175}V_{0.165}$	1.79	5.1	bcc	
$Nb_{0.2}Ta_{0.2}Ti_{0.2}Zr_{0.2}Fe_{0.2}$	1.61	6.9	bcc	
$Nb_{0.2}Ta_{0.2}Ti_{0.2}Zr_{0.2}Ge_{0.2}$	1.61	8.4	bcc	
$Nb_{0.2}Ta_{0.2}Ti_{0.2}Zr_{0.2}Si_{0.2}V_{0.2}$	1.79	4.3	bcc	
$Nb_{0.2}Ta_{0.2}Ti_{0.2}Zr_{0.2}Si_{0.2}Ge_{0.2}$	1.79	7.4	bcc	
$(TaNb)_{0.31}(TiUHf)_{0.69}$	1.59	3.2	bcc	[26]

Table 1.
List of HEA superconductors and HEA-type superconducting compounds; composition, mixing entropy, T_c, and structural type are summarized.

physical properties in various crystal structures, we could identify the merit of HEA states in those compounds. In section 2, we review the material synthesis, crystal structure, and physical properties of newly synthesized HEA-type compounds.

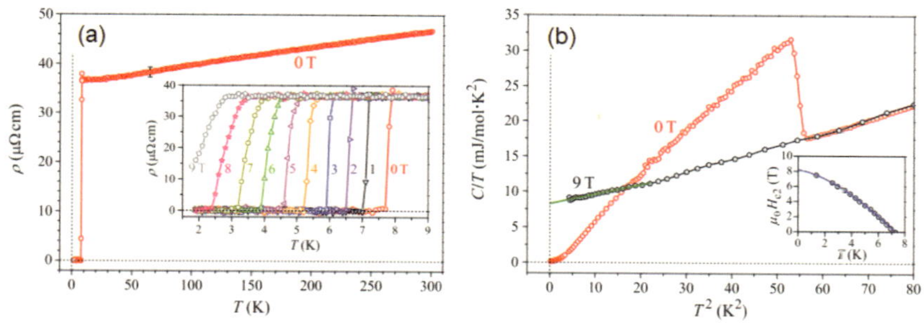

Figure 2.
(a) Temperature dependence of electrical resistivity (ρ) of $Ta_{0.34}Nb_{0.33}Hf_{0.08}Zr_{0.14}Ti_{0.11}$. (b) T^2 dependence of C/T, where C is specific heat of $Ta_{0.34}Nb_{0.33}Hf_{0.08}Zr_{0.14}Ti_{0.11}$. The figures were reproduced under permission by the authors of Ref. [18] (DOI: 10.1103/PhysRevLett.113.107001) and APS. Copyright 2014 by American Physical Society.

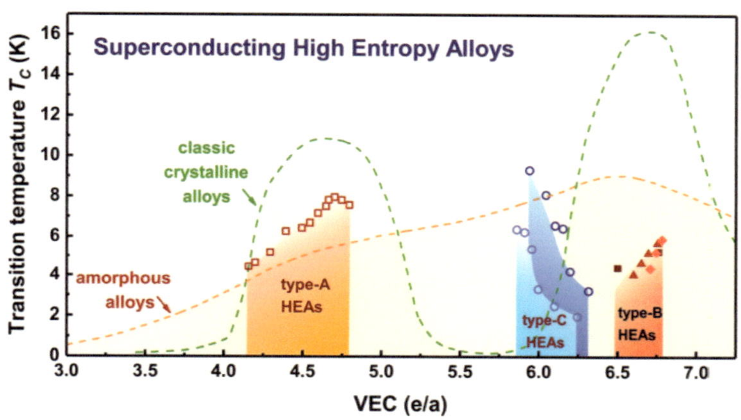

Figure 3.
Valence electron count (VEC) dependence of T_c for classic crystalline alloys, amorphous alloys, and HEAs. The figure was reproduced under permission by the authors of Ref. [15] (10.1103/PhysRevMaterials.3.090301) and APS. Copyright 2019 by American Physical Society.

Composition	$\Delta S_{mix}/R$ (site1)	$\Delta S_{mix}/R$ (site2)	$\Delta S_{mix}/R$ (Total)	T_c (K)	Structure	Ref.
$(ScZrNbTa)_{0.65}(RhPd)_{0.35}$	1.18	0.61	1.79	9.3	CsCl $Pm3m$	[29]
$(ScZrNb)_{0.63}(RhPd)_{0.37}$	1.16	0.62	1.79	7.5		
$(ScZrNb)_{0.62}(RhPd)_{0.38}$	1.16	0.63	1.79	6.4		
$(ScZrNb)_{0.60}(RhPd)_{0.40}$	1.14	0.64	1.78	3.9		
$(Nb_{0.11}Re_{0.56})(HfZrTi)_{0.33}$	0.73	0.57	1.30	4.4	hcp	[30]
$(Ag_{0.2}In_{0.2}Sn_{0.2}Pb_{0.2}Bi_{0.2})Te$	1.61	0	1.61	2.8	NaCl $Fm\text{-}3m$	[31]
$(Ag_{0.20}Cd_{0.20}Sn_{0.20}Sb_{0.15}Pb_{0.20})Te_{1.05}$	1.60	0	1.60	1.2	NaCl $Fm\text{-}3m$	[32]
$(Ag_{0.24}In_{0.22}Sn_{0.18}Sb_{0.14}Pb_{0.19})Te_{1.03}$	1.59	0	1.59	1.4		
$(Ag_{0.22}Cd_{0.22}In_{0.23}Sn_{0.17}Sb_{0.14})Te_{1.02}$	1.59	0	1.59	0.7		
$(Ag_{0.19}Cd_{0.19}Sn_{0.20}Pb_{0.18}Bi_{0.21})Te_{1.03}$	1.61	0	1.61	1.0		
$(Ag_{0.21}Cd_{0.19}In_{0.25}Pb_{0.16}Bi_{0.18})Te_{1.00}$	1.60	0	1.60	1.0		
$(Ag_{0.21}Cd_{0.21}In_{0.24}Sn_{0.19}Bi_{0.19})Te_{0.97}$	1.61	0	1.61	1.0		

Composition	$\Delta S_{mix}/R$ (site1)	$\Delta S_{mix}/R$ (site2)	$\Delta S_{mix}/R$ (Total)	T_c (K)	Structure	Ref.
$(Ag_{0.24}In_{0.22}Pb_{0.27}Bi_{0.26})Te_{1.02}$	1.37	0.00	1.37	2.7	NaCl *Fm*-3*m*	[33]
$(Ag_{0.29}In_{0.26}Pb_{0.22}Bi_{0.24})(Te_{0.78}Se_{0.20})$	1.38	0.51	1.89	2.5		
$(Ag_{0.34}In_{0.15}Pb_{0.24}Bi_{0.29})(Te_{0.65}Se_{0.34})$	1.35	0.65	2.00	2.0		
$(Co_{0.2}Ni_{0.1}Cu_{0.1}Rh_{0.3}Ir_{0.3})Zr_2$	1.47	0	1.47	7.8	$CuAl_2$ *I4/mcm*	[34]
$(Fe_{0.093}Co_{0.194}Ni_{0.113}Rh_{0.271}Ir_{0.329})Zr_2$	1.50	0	1.50	7.8	$CuAl_2$ *I4/mcm*	[35]
$(Fe_{0.108}Co_{0.297}Ni_{0.202}Rh_{0.073}Ir_{0.320})Zr_2$	1.48	0	1.48	6.7		
$(Fe_{0.190}Co_{0.190}Ni_{0.200}Rh_{0.212}Ir_{0.208})Zr_2$	1.61	0	1.61	5.4		
$(Fe_{0.293}Co_{0.190}Ni_{0.300}Rh_{0.093}Ir_{0.124})Zr_2$	1.52	0	1.52	4.8		
$(Y_{0.28}Nd_{0.16}Sm_{0.18}Eu_{0.18}Gd_{0.20})Ba_2Cu_3O_{7-d}$	1.59	0	1.59	93.0	Layered *Pmmm*	[36]
$(Y_{0.18}La_{0.24}Nd_{0.14}Sm_{0.14}Eu_{0.15}Gd_{0.15})Ba_2Cu_3O_{7-d}$	1.77	0	1.77	93.0		
$(La_{0.2}Ce_{0.2}Pr_{0.2}Nd_{0.2}Sm_{0.2})O_{0.5}F_{0.5}BiS_2$	1.61	0.69	2.30	4.3	Layered *P4/nmm*	[37, 38]
$(La_{0.3}Ce_{0.3}Pr_{0.2}Nd_{0.1}Sm_{0.1})O_{0.5}F_{0.5}BiS_2$	1.50	0.69	2.20	3.4	Layered *P4/nmm*	[37]
$(La_{0.1}Ce_{0.1}Pr_{0.3}Nd_{0.3}Sm_{0.2})O_{0.5}F_{0.5}BiS_2$	1.50	0.69	2.20	4.7		
$(La_{0.1}Ce_{0.1}Pr_{0.2}Nd_{0.3}Sm_{0.3})O_{0.5}F_{0.5}BiS_2$	1.50	0.69	2.20	4.9		
$(La_{0.28}Ce_{0.32}Pr_{0.21}Nd_{0.09}Sm_{0.10})BiS_2$	1.50	0	1.50	3.4	Layered *P4/nmm*	[39]
$(La_{0.10}Ce_{0.29}Pr_{0.33}Nd_{0.19}Sm_{0.09})BiS_2$	1.49	0	1.49	4.3		
$(La_{0.23}Ce_{0.21}Pr_{0.19}Nd_{0.19}Sm_{0.17})BiS_2$	1.60	0	1.60	3.3		
$(La_{0.09}Ce_{0.29}Pr_{0.12}Nd_{0.21}Sm_{0.29})BiS_2$	1.52	0	1.52	4.6		

Table 2.
List of HEA-type superconducting compounds; composition, mixing entropy (site-1, site-2, total), T_c, and structural type are summarized.

2. NaCl-type metal chalcogenides *MCh*

2.1 Metal tellurides *M*Te

The NaCl-type metal telluride family is one of the hot systems because it includes thermoelectric PbTe [40] and a topological crystalline insulator SnTe [41]. For superconducting tellurides, high-pressure synthesis was used to stabilize the NaCl-type structure [42–44]. For example, the low-pressure phase of InTe has a TlSe-type structure, but the high-pressure phase of InTe has a NaCl-type structure. The high-pressure phase can be obtained by high-pressure synthesis [43, 44]. Motivated by these facts, we tried to synthesize HEA-type tellurides *M*Te where the *M* site is in the HEA state (see **Figure 4(b)** for crystal structure) by high-pressure synthesis.

Figure 5(a) shows the temperature dependence of electrical resistivity for $AgInSnPbBiTe_5$, in which the *M* site is evenly occupied by Ag, In, Sn, Pb, and Bi (five metals) [31]. Very small *RRR* was observed, which is a similar trend to that in HEA superconductors [18]. In addition, four different *M*Te (*M*: Ag, In, Cd, Sn, Sb, Pb, Bi) superconductors with a HEA-type site has been obtained [32]. Interestingly, there is a correlation between the lattice constant and T_c in HEA-type *M*Te. In **Figure 5(b)**, the data for typical *M*Te superconductors are plotted. It is found that the trend that T_c increases with increasing lattice constant is common among the plotted *M*Te. The T_cs of HEA-type are, however, lower than those of low-entropy tellurides, such as InTe and (In,Sn)Te. Therefore, the introduction of the HEA-type *M* site is found to negatively work for T_c in MTe. This negative effect would be due to the direct effect of strong disorder to the *M*-Te bonding states and hence electronic states.

2.2 Hybrid high-entropy alloying in *MCh*

The Te site of *M*Te can be substituted by S and Se. The flexibility of both *M* and Te sites to element substitution enables us to design "*hybrid HEA*", in which both sites are alloyed [33]. **Figure 6(a)** shows the X-ray diffraction patterns for

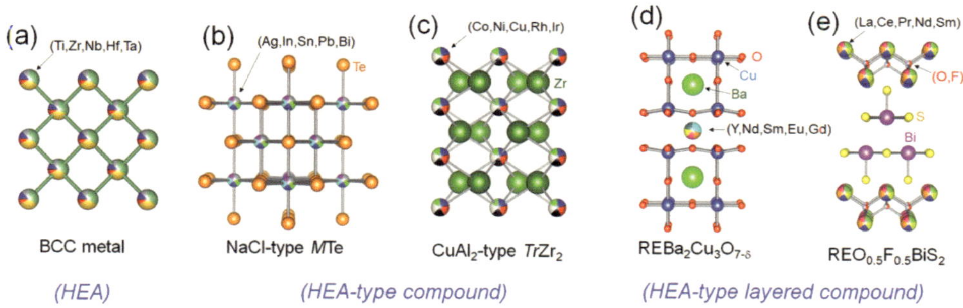

Figure 4.
Crystal structure images of conventional HEA (a) and HEA-type compounds (b–e).

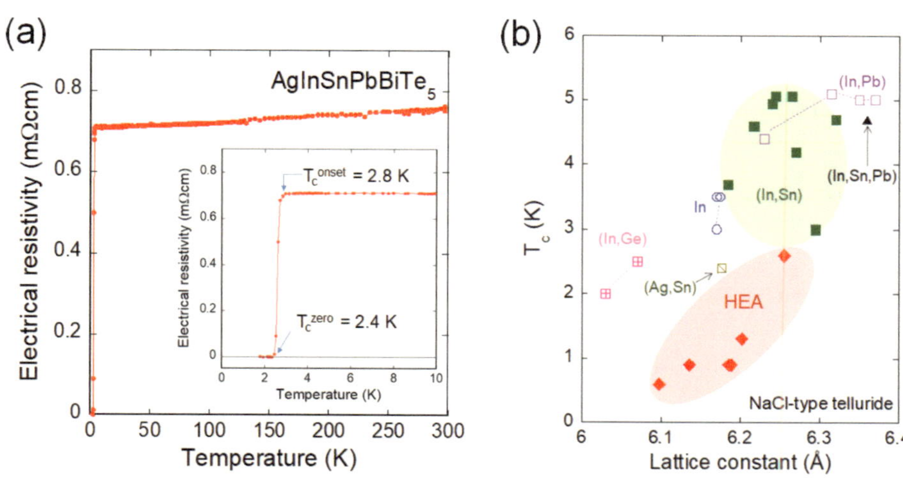

Figure 5.
(a) Temperature dependence of the electrical resistivity of $AgInSnPbBiTe_5$. Original data has been published in Ref. [31]. Copyright 2019 by the Physical Society of Japan. (b) Relationship between lattice constant and T_c for metal tellurides including HEA-type tellurides. The elements written in the figure indicate compositions of the M site. Original data has been published in Ref. [32]. Copyright 2020 by IOP.

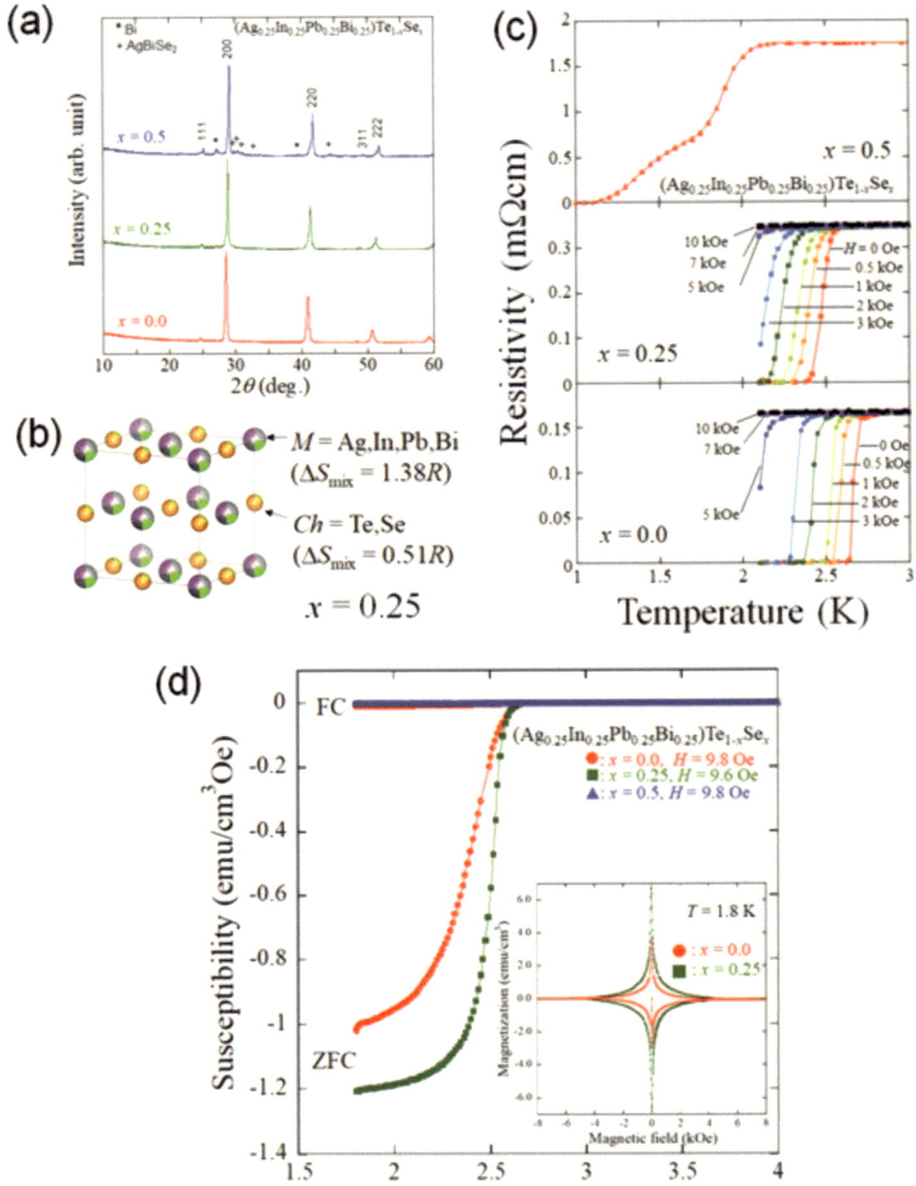

Figure 6.
(a) X-ray diffraction patterns for hybrid-HEA-type $(Ag_{0.25}In_{0.25}Pb_{0.25}Bi_{0.25})Te_{1}\text{-}xSe_{x}$. (b) Schematic image of the crystal structure and mixing entropy for x = 0.25 (c) temperature dependences of electrical resistivity $(Ag_{0.25}In_{0.25}Pb_{0.25}Bi_{0.25})Te_{1}\text{-}xSe_{x}$ under magnetic fields. (d) Temperature dependence of magnetic susceptibility. The inset shows the magnetization-field curves. Original data has been published in Ref. [33]. Copyright 2020 by the Royal Society of Chemistry.

$(Ag_{0.25}In_{0.25}Pb_{0.25}Bi_{0.25})Te_{1-x}Se_{x}$. Notably, mixing many elements at two sites does not result in phase separation, and a single-phase sample was obtained for x = 0.25, while small impurity phases were detected for x = 0.5. For x = 0.25, as displayed in **Figure 6(b)**, the ΔS_{mix} for the *M* and *Ch* sites are 1.38*R* and 0.51*R*, and the total ΔS_{mix} reaches 1.89*R*. Furthermore, the total ΔS_{mix} for x = 0.5 reaches 2.00*R*. These ΔS_{mix} values are clearly higher than that for HEAs (**Table 1**). Therefore, alloying at two or more sites (hybrid high-entropy alloying) results in very high total ΔS_{mix}.

The superconducting properties for $(Ag_{0.25}In_{0.25}Pb_{0.25}Bi_{0.25})Te_{1-x}Se_x$ are shown in **Figure 6(c)**. A superconducting transition with $T_c > 2$ K was observed for $x = 0$ and 0.25. By focusing on these two phases, interesting trend was found. Although the T_c for $x = 0.25$ is lower than that for $x = 0$, the suppression of T_c for $x = 0.25$ under magnetic fields is clearly smaller than that for $x = 0$. From the estimation of upper critical field (H_{c2}), it was found that the H_{c2} (0 K) for $x = 0.25$ is higher than that for $x = 0$. In addition, from the measurements of the magnetization-field loop, it was confirmed that the critical current density (J_c) at 1.8 K for $x = 0.25$ is larger than that for $x = 0$. These results suggest that an increase in ΔS_{mix} may be useful to improve H_{c2} and/or J_c characteristics of superconductors if the problem on the suppression of T_c could be solved.

3. $CuAl_2$-type $TrZr_2$

As shown in the last section, high-entropy alloying of a compound, which possesses two or more crystallographic sites, is a route to develop new superconductors with a high mixing entropy. In addition, in HEA-type compounds, not only the site entropy but also the entropy of chemical bonding states should be higher than normal alloys or compounds. To discover new HEA-type superconductors, the use of material database is quite useful. SuperCon (NIMS database) [45] is a database of superconductors and contain information of composition, structural type, T_c, and reference of the material. To achieve the material design of HEA-type superconductors, we have to find a system in which compositional variation is rich within the same structural type, and superconducting transition has been observed. Herein, we introduce an example of material design for new HEA-type superconductors.

$TrZr_2$ (T_r: transition metal) with a tetragonal $CuAl_2$-type structure (**Figure 4(c)**) is a superconducting system. For Tr = Fe, Co, Ni, Rh, Ir, superconductivity was reported. Among them, $CoZr_2$, $RhZr_2$, and $IrZr_2$ exhibits a higher T_c of 5.5, 11.3, and 7.5 K, respectively [46]. Furthermore, the Tr site can be partially substituted by Sc, Cu, Ga, Pd, and Ta [34, 45]. These facts suggest that the Tr site of $TrZr_2$ can be modified into a HEA-type site. We synthesized $Co_{0.2}Ni_{0.1}Cu_{0.1}Rh_{0.3}Ir_{0.3}Zr_2$ by arc melting and observed superconductivity with a T_c of 8 K [34]. The Tr site of $Co_{0.2}Ni_{0.1}Cu_{0.1}Rh_{0.3}Ir_{0.3}Zr_2$ contains five transition metals, and the ΔS_{mix} for the Tr site is about 1.5R. In addition, (Fe,Co,Ni,Rh,Ir)Zr_2 superconductors were also synthesized and exhibited superconductivity [35]. Interestingly, the resulting T_c in HEA-type phases was very close to the weighted-average T_c of pure $TrZr_2$ systems. Although the origin of the unexpected behavior is still unclear, the effects of disordering by the HEA-type site to superconducting properties seem very limited. The difference in the HEA effects between MCh and $TrZr_2$ may be caused by the different structural complexity.

As shown in **Figure 7(a)**, the temperature dependence of resistivity exhibits a relatively large RRR as compared to other HEA-type superconductors. However, a large RRR of ~30 was reported for $CoZr_2$ single crystals [47], which is clearly larger than that for HEA-type $TrZr_2$. Therefore, the HEA effects (disordering effects) to transport properties would be a common trend. **Figure 7(b)** shows the temperature dependences of the specific heat for $Co_{0.2}Ni_{0.1}Cu_{0.1}Rh_{0.3}Ir_{0.3}Zr_2$ in the form of C/T. Although a clear jump of C/T at the T_c is seen, the transition is broader as compared to the case of $CoZr_2$. The trend of broad transition in specific heat at T_c was not observed in a HEA superconductor but observed in HEA-type compounds. Because the superconducting transitions observed in resistivity and magnetization were sharp, the broad transition in the specific heat would suggest microscopic local phase separation in HEA-type superconducting compounds.

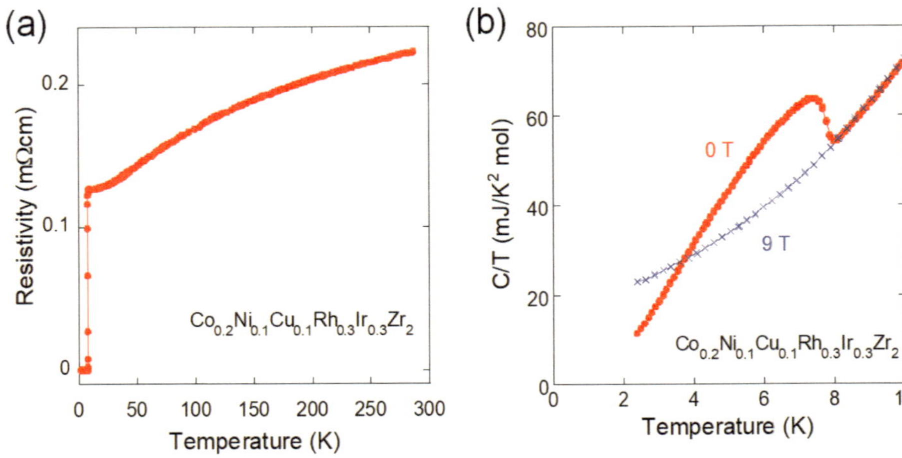

Figure 7.
(a) Temperature dependence of resistivity for $Co_{0.2}Ni_{0.1}Cu_{0.1}Rh_{0.3}Ir_{0.3}Zr_2$. (b) Temperature dependence of specific heat (C/T) for $Co_{0.2}Ni_{0.1}Cu_{0.1}Rh_{0.3}Ir_{0.3}Zr_2$ at 0 and 9 T. Original data has been published in Refs. [34] and [35]. Copyright 2020 by Taylor & Francis.

4. Cuprate (high-T_c) superconductors $REBa_2Cu_3O_{7-d}$

As mentioned in introduction, cuprates (Cu oxides) have been extensively studied in the fields of science and engineering because of its high T_c. Among them, $REBa_2Cu_3O_{7-d}$ (RE123) system [5] is one of practical materials for superconductivity

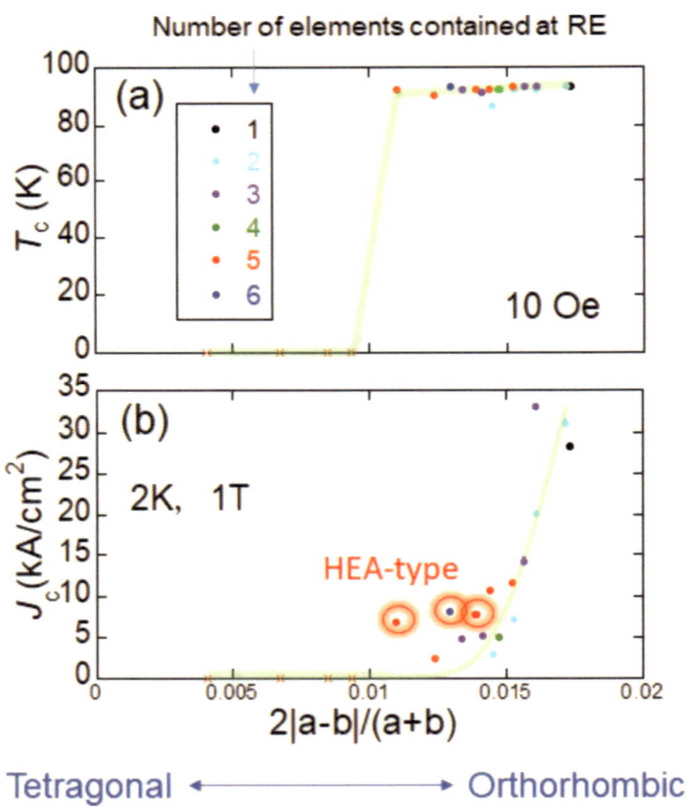

Figure 8.
Orthorhombicity parameter (2|a-b|/(a + b)) dependences of (a) T_c and (b) magnetic J_c (T = 2 K, B = 1 T) for $REBa_2Cu_3O_{7-d}$. Original data has been published in Ref. [36]. Copyright 2020 by Elsevier.

application. In addition, a high J_c was reported in *RE*123 samples with three elements at the *RE* site [48]. Motivated by the fact, we synthesized polycrystalline samples of $REBa_2Cu_3O_{7-d}$ with different ΔS_{mix} for the *RE* site [36] by standard solid-state reaction in air. In the study, two-step annealing was performed to optimize oxygen content (*d*) because oxygen content affect crystal structure (orthorhombicity) and superconducting properties of $REBa_2Cu_3O_{7-d}$. A high superconducting property is generally achieved in an orthorhombic phase in the system, we estimated the orthorhombicity parameter (*OP*), which is given by $2|a-b|/(a+b)$, and plotted the estimated T_c and magnetic J_c (T = 2 K, B = 1 T) for $REBa_2Cu_3O_{7-d}$ as a function of *OP* as shown in **Figure 8**. Note that the data points are colored according to the number of elements contained at the RE site. From the plot, it is found that Tc does not show a remarkable correlation with ΔS_{mix} (RE site) and exhibits a clear correlation with *OP*. J_c also exhibits a trend of improvement with increasing *OP*. These facts suggest that disorder introduced by high-entropy alloying at the *RE* site (see **Figure 4(d)**) does not largely affect superconducting properties of $REBa_2Cu_3O_{7-d}$, which is a two-dimensional layered compound. Because the trend is clearly different to that observed for cubic (NaCl-type) tellurides with a HEA-type site, crystal-structure dimensionality is a key factor to how the introduction of HEA-type site affects superconducting properties in compounds.

In **Figure 8(b)**, we found three data points for HEA-type samples show a J_c larger than that for the other low-entropy samples at *OP* = 0.01–0.015. Although it has not been fully clarified whether the slightly large J_cs in the HEA-type samples are caused by high-entropy alloying or not, the effect of high-entropy alloying for cuprates should be further studied to find the way to improve practical performance of cuprate superconductors.

5. BiS_2-based layered superconductors *RE*(O,F)BiS_2

BiS_2-based superconductor family is one of the layered superconductor families and was discover in 2012 [49–51]. The crystal structure is composed of alternate stacks of a conducting BiS_2 bilayer and a blocking layer (for example, a *RE*O layer), which is similar to that of high-T_c systems. Furthermore, unconventional superconductivity has been proposed from theoretical and experimental studies on the BiS_2-based compounds [51].

A typical BiS_2-based system is $REOBiS_2$ (see **Figure 4(e)**). Because non-doped $REOBiS_2$ is a semiconductor, electron carrier doping is required to induce metallicity [50]. For *RE* = La, a superconducting transition was observed at 2.5 K after electron doping through partial substitution of O by F in $LaO_{0.5}F_{0.5}BiS_2$. However, the superconductivity states in La(O,F)BiS_2 is not bulk in nature. This is due to the presence of the local disorder due to Bi lone pairs, and the local disorder could be suppressed by in-plane chemical pressure effects [52–54]. In-plane chemical pressure can be generated by *RE*-site substitution by smaller *RE* ions or Se substitution for the S site. By increasing in-plane chemical pressure, local disorder is suppressed, and bulk superconductivity is induced [52, 53]. Therefore, one can say that *RE*(O,F)BiS_2 is a useful system to investigate the effect of structural modification on local crystal structure and superconducting properties. This suggests that the investigation of the effects of introduction of a HEA site in *RE*(O,F)BiS_2 would provide us with key information about interlayer interaction through the HEA states.

The HEA-type samples of $REO_{0.5}F_{0.5}BiS_2$ with *RE* = La, Ce, Pr, Nd, Sm were synthesized by solid-state reaction in an evacuated quartz tube [37]. **Figure 9(a)** and **(b)** show superconducting properties of four different HEA-type $REO_{0.5}F_{0.5}BiS_2$ samples. The T_c varies depending on the *RE*-site composition. As shown in **Figure 9(c)**, these

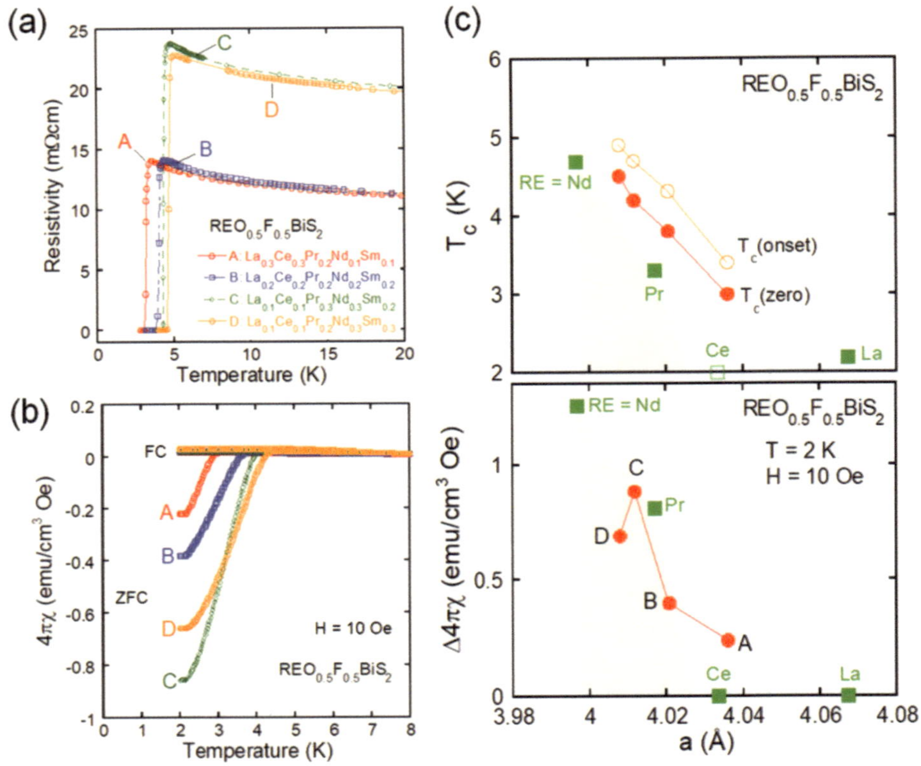

Figure 9.
(a, b) Temperature dependences of (a) electrical resistivity and (b) magnetic susceptibility for HEA-type $REO_{0.5}F_{0.5}BiS_2$. (c) Lattice constant a dependences of T_c and $\Delta 4\pi\chi$ (T = 2 K). Original data has been published in Ref. [37]. Copyright 2020 by IOP.

four samples have different lattice constants a, which is corresponding to different in-plane chemical pressures. Therefore, the variation of T_c can be understood with the in-plane chemical pressure scenario. When focusing on the samples with the same lattice constant a and different mixing entropy, we find that the superconducting properties (T_c and $\Delta 4\pi\chi$) of the HEA-type sample are higher than those of low-entropy samples (**Figures 9(c)** and **(d)**). The facts indicate that an increase in ΔS_{mix} for the *RE* site may positively work in improving superconducting properties. Therefore, we systematically prepared REO$_{0.5}$F$_{0.5}$BiS$_2$ samples with the same lattice constant a and different ΔS_{mix} to investigate the interlayer interaction [38].

As summarized in **Figure 10(a)**, we succeeded in preparation of a set of five REO$_{0.5}$F$_{0.5}$BiS$_2$ samples with almost the same a and systematically varied ΔS_{mix}. Because the lattice constant a nearly corresponds to the degree of the in-plane chemical pressure, the set of samples have almost similar in-plane chemical pressure. Therefore, we could detect the effects of the increase in ΔS_{mix} to superconducting properties and local structural disorder [38]. Through magnetic susceptibility measurements, we confirmed that shielding volume fraction increases with increasing ΔS_{mix} while the T_c does not change (**Figure 10(c)** and **(d)**). To understand the origin of the improvement of superconducting properties, local structure was analyzed using synchrotron X-ray diffraction. As summarized in **Figure 10(b)**, suppression of in-plane structural disorder, which was detected by anisotropic displacement parameter for the in-plane direction (U_{11}), was achieved by the HEA effect at the *RE* site. Similar trend was observed in another set of REO$_{0.5}$F$_{0.5}$BiS$_2$ samples with larger lattice constant (comparable to that of $CeO_{0.5}F_{0.5}BiS_2$). The results propose that high-entropy alloying at the blocking layer would modify

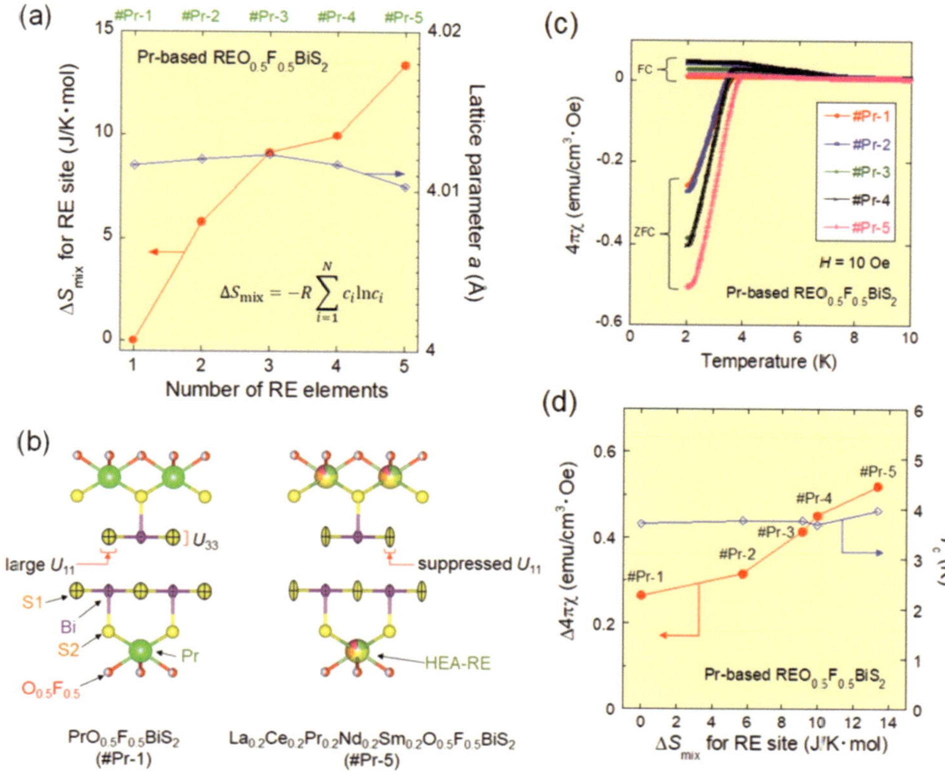

Figure 10.
(a) Mixing entropy (ΔS_{mix}) and lattice constant a are plotted as a function of the number of RE elements contained in the sample. (b) Schematic images of local structural disorder for low-entropy and HEA-type $REO_{0.5}F_{0.5}BiS_2$. (c) Temperature dependences of $4\pi\chi$. (d) ΔS_{mix} dependence of $\Delta 4\pi\chi$ (T = 2 K). Original data has been published in Ref. [38]. Copyright 2020 by Elsevier.

the local structure at the conducting layer in BiS_2-based layered compounds. This effect should be a new strategy to develop novel functional materials with a layered structure.

6. Conclusion

In this chapter, we reviewed the works on HEA superconductors and HEA-type superconducting compounds, which have been developed applying the HEA concept in simple alloys to more complicated compounds. By introducing a HEA-type site (or alloying multi sites) in a compound, a high mixing entropy can be achieved. HEA-type compounds would have a high mixing entropy at the alloyed site and a highly (randomly) disordered bonding states. For a three-dimensional system, such as NaCl-type chalcogenides, introduction of the HEA-type *M* site in *MCh* resulted in suppression of T_c, but, at the same time, slight enhancements of H_{c2} and J_c were observed. For tetragonal ($CuAl_2$-type) $TrZr_2$, the introduction of the HEA-type site does not suppress T_c. Furthermore, for layered systems, such as $REBa_2Cu_3O_{7-d}$ cuprates and BiS_2-based $RE_{0.5}F_{0.5}BiS_2$, high-entropy alloying rather improves their superconducting properties. Through investigations of the HEA effects to superconducting and structural properties for HEA-type compounds with different structural types, we conclude that the HEA effects are highly depending on the structural dimensionality. Hence, to effectively use the HEA effect to improve superconducting properties of compounds, target structure should be lower-symmetric, a layered structure or a one-dimensional structure. To obtain further knowledge about the presence/absence of local phase separations, pinning characteristics, and the HEA effect to superconducting pairing states, further investigations on HEA-type superconducting compounds using various probes are required.

Acknowledgements

The author thanks R. Sogabe, Md. R. Kasem, Y. Shukunami, M. Katsuno, K. Hoshi, R. Jha, Y. Goto, T. D. Matsuda, and O. Miura who have contributed on the works reviewed here. The author would like to thank all the coauthors of the works. The works reviewed here were supported by JSPS-KAKENHI (Nos.: 16H04493, 15H05886, 18KK0076), JST-CREST (Nos.: JPMJCR16Q6, JPMJCR20Q4), and Tokyo Metropolitan Government Advanced Research (No.: H31-1).

Author details

Yoshikazu Mizuguchi* and Aichi Yamashita
Department of Physics, Tokyo Metropolitan University, Hachioji, Japan

*Address all correspondence to: mizugu@tmu.ac.jp

References

[1] Onnes H. K. The Superconductivity of Mercury. Comm. Phys. Lab. Univ., Leiden, 1911; 122-124.

[2] Adir Moyses Luiz (2011). Room Temperature Superconductivity, Superconductivity - Theory and Applications, Dr. Adir Luiz (Ed.), ISBN: 978-953-307-151-0, InTech, Available from: https://www.intechopen.com/books/superconductivity-theory-and-applications/room-temperature-superconductivity

[3] Bednortz J. G., Müller K. A. Possible High T_c Superconductivity in the Ba-La-Cu-O System. Z. Phys. B - Condens. Matter 1986;64 189-193.

[4] Takagi H., Uchida S., Kitazawa K., Tanaka S. High-T_c Superconductivity of La-Ba-Cu Oxides. II. –Specification of the Superconducting Phase. Jpn. J. Appl. Phys. 1987;26 L123–L124.

[5] Wu M. K., Ashburn J. R., Torng C. J., Hor P. H., Meng R. L., Gao L., Huang Z. J., Wang Y. Q., Chu C. W. Superconductivity at 93 K in a new mixed-phase Y-Ba-Cu-O compound system at ambient pressure. Phys. Rev. Lett. 1987;58 908-910.

[6] Schilling A., Cantoni M., Guo J. D., Ott H. R. Superconductivity above 130 K in the Hg-Ba-Ca-Cu-O system. Nature 1993; 363 56-58.

[7] Nagamatsu J., Nakagawa N. Muranaka T., Zenitani Y., Akimitsu J. Superconductivity at 39 K in magnesium diboride. Nature 2001;410 63-64.

[8] Kamihara Y., Watanabe T., Hirano M., Hosono H. Iron-Based Layered Superconductor La[$O_{1-x}F_x$]FeAs (x = 0.05-0.12) with Tc = 26 K. J. Am. Chem. Soc. 2008;130 3296-3297.

[9] Ren Z. A., Lu W., Yang J., Yi W., Shen X. L., Li Z. C., Che G. C., Dong X. L., Sun L. L., Z. F., Z. Z. X. Superconductivity at 55 K in Iron-Based F-Doped Layered Quaternary Compound Sm[$O_{1-x}F_x$]FeAs. Chin. Phys. Lett. 2008;25 2215-2216.

[10] Norman M. R. The Challenge of Unconventional Superconductivity. Science 2011;332 196-200.

[11] Drozdov A. P., Eremets M. I., Troyan I. A., Ksenofontov V., Shylin S. I. Conventional superconductivity at 203 kelvin at high pressures in the sulfur hydride system. Nature 2015;525, 73-76.

[12] Drozdov A. P., Kong P. P., Minkov V. S., Besedin S. P., Kuzovnikov M. A., Mozaffari S., Balicas L., Balakirev F. F., Graf D. E., Prakapenka V. B., Greenberg E., Knyazev D. A., Tkacz M., Eremets M. I. Superconductivity at 250 K in lanthanum hydride under high pressures. Nature 2019;569 528-531.

[13] Somayazulu M., Ahart M., Mishra A. K., Geballe Z. M., Baldini M., Meng Y., Struzhkin V. V., Hemley R. J. Evidence for Superconductivity above 260 K in Lanthanum Superhydride at Megabar Pressures. Phys. Rev. Lett. 2019;122 027001(1-6).

[14] Snider E., Dasenbrock-Gammon N., McBride R., Debessai M., Vindana H., Vencatasamy K., Lawler K. V., Salamat A., Dias R. P. Room-temperature superconductivity in a carbonaceous sulfur hydride. Nature 2020;586 373-377.

[15] Sun L., Cava R. J. High-entropy alloy superconductors: Status, opportunities, and challenges. Phys. Rev. Mater. 2019;3 090301(1-10).

[16] Yeh J. W., Chen S. K., Lin S. J., Gan J. Y., Chin T. S., Shun T. T., Tsau C. H., Chang S. Y. Nanostructured High-Entropy Alloys with Multiple Principal Elements: Novel Alloy Design Concepts

and Outcomes. Adv. Eng. Mater. 2004;6 299-303.

[17] Tsai M. H., Yeh J. W. High-Entropy Alloys: A Critical Review. Mater. Res. Lett. 2014;2 107-123.

[18] Koželj P., Vrtnik S., Jelen A., Jazbec S., Jagličić Z., Maiti S., Feuerbacher M., Steurer W., Dolinšek J. Discovery of a Superconducting High-Entropy Alloy. Phys, Rev. Lett. 2014;113 107001(1-5).

[19] Vrtnik S., Koželj P., Meden A., Maiti S., Steurer W., Feuerbacher M., Dolinšek J. Superconductivity in thermally annealed Ta-Nb-Hf-Zr-Ti high-entropy alloys. J. Alloy Compounds 2017;695 3530-3540.

[20] Guo J., Wang H., von Rohr F., Wang Z., Cai S., Zhou Y., Yang K., Li A., Jiang S., Wu Q., Cava R. J., Sun L. Robust zero resistance in a superconducting high-entropy alloy at pressures up to 190 GPa. Proc. Natl. Acad. Sci. U.S.A. 2017;114 13144-13147.

[21] von Rohr F., Cava R. J. Isoelectronic substitutions and aluminium alloying in the Ta-Nb-Hf-Zr-Ti high-entropy alloy superconductor. Phys. Rev. Mater. 2018;2 034801(1-7).

[22] Ishizu N., Kitagawa J. New high-entropy alloy superconductor $Hf_{21}Nb_{25}Ti_{15}V_{15}Zr_{24}$. Results in Phys. 2019;13 102275(1-2).

[23] Yuan Y., Wu Y., Luo H., Wang Z., Liang X., Yang Z., Wang H., Liu X., Lu Z. Superconducting $Ti_{15}Zr_{15}Nb_{35}Ta_{35}$ High-Entropy Alloy With Intermediate Electron-Phonon Coupling. Front. Mater. 2018;5 72(1-6).

[24] Stolze K., Cevallos F. A., Kong T., Cava R. J. High-entropy alloy superconductors on an α-Mn lattice. J. Mater. Chem. C 2018;6 10441-10449.

[25] Wu K. J. Chen S. K., Wu J. M. Superconducting in Equal Molar NbTaTiZr-Based High-Entropy Alloys. Natural Sci. 2018;10 110-124.

[26] Nelson W. L., Chemey A. T., Hertz M., Choi E., Graf D. E., Latturner S., Albrecht-Schmitt T. E., WeiK., BaumbachR.E.Superconductivity in a uranium containing high entropy alloy. Sci. Rep. 2020:10 4717(1-8).

[27] von Rohr F., Winiarski M. J., Tao J., Klimczuk T., Cava R. J. Effect of electron count and chemical complexity in the Ta-Nb-Hf-Zr-Ti high-entropy alloy superconductor. Proc. Natl. Acad. Sci. U.S.A. 2016;113 7144-7150.

[28] Guo J., Lin G., Cai S., Xi C., Zhang C., Sun W., Wang Q., Yang K., Li A., Wu Q., Zhang Y., Xiang T., Cava R. J., Sun L. Record-High Superconductivity in Niobium–Titanium Alloy. Adv. Mater. 2019;31 1807240(1-5).

[29] Stolze K., Tao J., von Rohr F., Kong T., Cava R. J. Sc–Zr–Nb–Rh–Pd and Sc–Zr–Nb–Ta–Rh–Pd High-Entropy Alloy Superconductors on a CsCl-Type Lattice. Chem. Mater. 2018;30 906-914.

[30] Marik S., Varghese M., Sajilesh K. P., Singh D., Singh R. P. Superconductivity in a new hexagonal high-entropy alloy. Phys. Rev. Mater. 2019;3 060602(1-6).

[31] Mizuguchi Y. Superconductivity in High-Entropy-Alloy Telluride $AgInSnPbBiTe_5$. J. Phys. Soc. Jpn. 2019;88 124708(1-5).

[32] Kasem Md. R., Hoshi K., Jha R., Katsuno M., Yamashita A., Goto Y., Matsuda T. D., Aoki Y., Mizuguchi Y. Superconducting properties of high-entropy-alloy tellurides M-Te (M: Ag, In, Cd, Sn, Sb, Pb, Bi) with a NaCl-type structure. Appl. Phys. Express 2020;13 033001(1-4).

[33] Yamashita A., Jha R., Goto Y., Matsuda T. D., Aoki Y., Mizuguchi Y. An efficient way of increasing the total entropy of mixing in high-entropy-alloy

compounds: a case of NaCl-type (Ag,In,Pb,Bi)$Te_{1-x}Se_x$ (x = 0.0, 0.25, 0.5) superconductors. Dalton Trans. 2020;49 9118-9122.

[34] Mizuguchi Y., Kasem Md. R., Matsuda T. D. Superconductivity in $CuAl_2$-type $Co_{0.2}Ni_{0.1}Cu_{0.1}Rh_{0.3}Ir_{0.3}Zr_2$ with a high-entropy-alloy transition metal site. Mater. Res. Lett. 2021;9 141-147.

[35] Kasem Md. R., Yamashita A., Goto Y., Matsuda T. D., Mizuguchi Y. Synthesis of high-entropy-alloy-type superconductors (Fe,Co,Ni,Rh,Ir)Zr_2 with tunable transition temperature. arXiv:2011.05590.

[36] Shukunami Y., Yamashita A., Goto Y., Mizuguchi Y. Synthesis of RE123 high-T_c superconductors with a high-entropy-alloy-type RE site. Physica C 2020;572 1353623(1-5).

[37] Sogabe R., Goto Y., Mizuguchi Y. Superconductivity in $REO_{0.5}F_{0.5}BiS_2$ with high-entropy-alloy-type blocking layers. Appl. Phys. Express 2018;11 053102(1-5).

[38] Sogabe R., Goto Y., Abe T., Moriyoshi C., Kuroiwa Y., Miura A., Tadanaga K., Mizuguchi Y. Improvement of superconducting properties by high mixing entropy at blocking layers in BiS_2-based superconductor $REO_{0.5}F_{0.5}BiS_2$. Solid State Commun. 2019;295 43-49.

[39] Fujita Y., Kinami K., Hanada Y., Nagao M., Miura A., Hirai S., Maruyama Y., Watauchi S., Takano Y., Tanaka I. Growth and Characterization of $ROBiS_2$ High-Entropy Superconducting Single Crystals. ACS Omega 2020;5 16819-16825.

[40] Heremans J. P., Jovovic V., Toberer E. S., Saramat A., Kurosaki K., Charoenphakdee A., Yamanaka S., Snyder G. J. Enhancement of Thermo-electric Efficiency in PbTe by Distortion of the Electronic Density of States. Science 2008;321 554-557.

[41] TanakaY.,RenZ.,SatoT.,NakayamaK., Souma S., Takahashi T., Segawa K., Ando Y. Experimental realization of a topological crystalline insulator in SnTe. Nat. Phys. 2012;8 800-803.

[42] Mizuguchi Y., Miura O. H igh-Pressure Synthesis and Superconductivity of Ag-Doped Topological Crystalline Insulator SnTe ($Sn_{1-x}Ag_xTe$ with x = 0-0.5). J. Phys. Soc. Jpn. 2016;85 053702(1-5).

[43] Kobayashi K., Ai Y., Jeschke H. O., Akimitsu J. Enhanced superconducting transition temperatures in the rocksalt-type superconductors $In_{1-x}Sn_xTe$ ($x \leq 0.5$). Phys. Rev. B 2018;97 104511(1-6).

[44] Katsuno M., Jha R., Hoshi K., Sogabe R., Goto Y., Mizuguchi Y. High-Pressure Synthesis and Superconducting Properties of NaCl-Type $In_{1-x}Pb_xTe$ (x = 0-0.8). Condens. Matter 2020;5 14(1-10).

[45] SuperCon (NIMS database): https://supercon.nims.go.jp/en/

[46] Fisk Z., Viswanathan R., Webb G. W. THE RELATION BETWEEN NORMAL STATE PROPERTIES AND T_c FOR SOME Zr_2X COMPOUNDS. Solid State Commun. 1974;15 1797-1799.

[47] Teruya A., Kakihana M., Takeuchi T., Aoki D., Honda F., Nakamura A., Haga Y., Matsubayashi K., Uwatoko Y., Harima H., Hedo M., Nakama T., Ōnuki Y. Superconducting and Fermi Surface Properties of Single Crystal Zr_2Co. J. Phys. Soc. Jpn. 2016; 85: 034706(1-10).

[48] Cai C., Holzapfel B., Hänisch J., Fernández L., Schultz L. High critical current density and its field dependence in mixed rare earth (Nd,Eu,Gd) $Ba_2Cu_3O_{7-\delta}$ thin films. Appl. Phys. Lett. 2004;84 377.

[49] Mizuguchi Y., Fujihisa H., Gotoh Y., Suzuki K., Usui H., Kuroki K., Demura S., Takano Y., Izawa H., Miura O. BiS_2-based layered superconductor $Bi_4O_4S_3$. Phys. Rev. B 2012;86 220510(1-5).

[50] Mizuguchi Y., Demura S., Deguchi K., Takano Y., Fujihisa H., Gotoh Y., Izawa H., Miura O. J. Phys. Soc. Jpn. 2012;81 114725(1-5).

[51] Mizuguchi Y. Material Development and Physical Properties of BiS_2-Based Layered Compounds. J. Phys. Soc. Jpn. 2019;88 041001 (1-17).

[52] Mizuguchi Y., Miura A., Kajitani J., Hiroi T., Miura O., Tadanaga K., Kumada N., Magome E., Moriyoshi C., Kuroiwa Y. In-plane chemical pressure essential for superconductivity in $BiCh_2$-based (Ch: S, Se) layered structure. Sci. Rep. 2015; 5 14968(1-8).

[53] Mizuguchi Y., Hoshi K., Goto Y., Miura A., Tadanaga K., Moriyoshi C., Kuroiwa Y. Evolution of Anisotropic Displacement Parameters and Superconductivity with Chemical Pressure in BiS_2-Based $REO_{0.5}F_{0.5}BiS_2$ (RE = La, Ce, Pr, and Nd). J. Phys. Soc. Jpn. 2018;87 023704 (1-4).

[54] Athauda A., Louca D. Nanoscale Atomic Distortions in the BiS_2 Superconductors: Ferrodistortive Sulfur Modes. J. Phys. Soc. Jpn. 2019;88 041004 (1-10).

Chapter 2

Materials Research on High-Entropy Alloy Superconductors

Jiro Kitagawa, Naoki Ishizu and Shusuke Hamamoto

Abstract

The first purpose of this chapter is materials research on face-centered-cubic (fcc) high-entropy alloy (HEA) superconductors, which have not yet been reported. We have investigated several Nb-containing multicomponent alloys. Although we succeeded in obtaining Nb-containing samples with the dominant fcc phases, no superconducting signals appeared in these samples down to 3 K. The microstructure analyses revealed that all samples were multi-phase, but the existence of several new Nb-containing HEA phases was confirmed in them. The second purpose is the report of materials research on the Mn_5Si_3-type HEA superconductors. This hexagonal structure offers various intermetallic compounds, which often undergo a superconducting state. The Mn_5Si_3-type HEA is classified into the multisite HEA, which possesses the high degree of freedom in the materials design and is good platform for studying exotic HEA superconductors. We have successfully found a single-phase Mn_5Si_3-type HEA, which, however, does not show a superconducting property down to 3 K. The attempt of controlling the valence electron count was not successful.

Keywords: high-entropy alloys, superconductivity, face-centered-cubic, niobium, Mn_5Si_3-type

1. Introduction

High-entropy alloys (HEAs) are a new class of materials and have attracted a great deal of attention [1, 2]. The concept of HEA was originally proposed for a face-centered-cubic (fcc), body-centered-cubic (bcc), or hexagonal-closed packing (hcp) structure. The most prominent feature of a HEA is that more than five elements, each having an atomic percentage between 5% and 35%, randomly occupy one crystallographic site (see also **Figure 1(a)**). This produces a large mixing entropy, and HEAs exhibit the combination of high yield strength and ductility [3], high strength at elevated temperatures [4], strong resistance to corrosion and oxidation [5], and so on. The high-entropy concept is extensively adapted in various materials such as oxides, chalcogenides, and halides [6, 7].

One of the novelties of HEAs is a cocktail effect, which indicates an enhancement of physical properties beyond the simple mixture of those of components. For example, several bcc HEAs show superior mechanical properties compared to conventional hard materials. Another example is found in magnetic spinel oxide $(Mg_{0.2}Co_{0.2}Ni_{0.2}Cu_{0.2}Zn_{0.2})Al_2O_4$. The high-entropy type spinel oxide interestingly

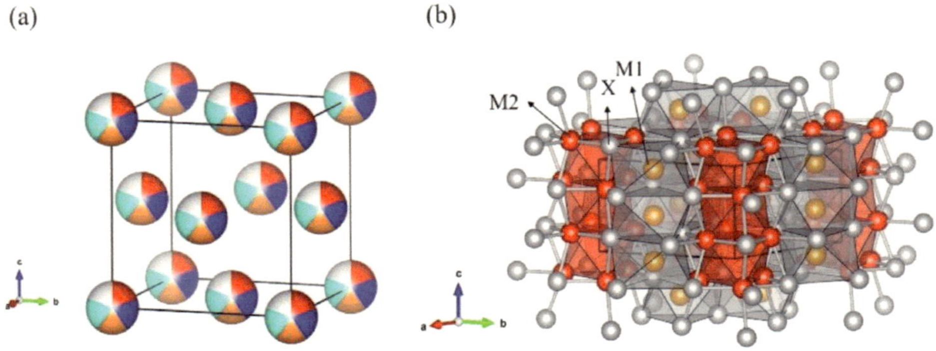

Figure 1.
Crystal structure of compounds with (a) fcc and (b) Mn_5Si_3-type structures. In (a), a multicolored ball means a random occupation by constituent elements.

shows enhanced magnetic frustration [8]. The cocktail effect is also reported in the structural stability of high-entropy-type materials. A γ-type disilicate structure is stable from room temperature to 1900°C in $(Gd_{1/6}Tb_{1/6}Dy_{1/6}Tm_{1/6}Yb_{1/6}Lu_{1/6})_2Si_2O_7$. The outstanding thermal stability is ascribed to the high-entropy state at the rare-earth site [9]. The other novelty of HEAs is the tuning of physical properties via the change of microstructure. The manufacturing process of HEAs considerably affects their microstructures, which are often deeply related to their physical properties. $Fe_{15}Co_{15}Ni_{20}Mn_{20}Cu_{30}$ shows a spinodal decomposition after the heat treatment [10]. The spinodally decomposed sample exhibits enhanced Curie temperature and magnetization compared to the homogenized single-phase sample. The tuning of magnetic properties is also reported in dual-phase HEAs [11–13].

One of the new research topics in HEA is the superconductivity found in 2014 [14]. Transition metal-based superconductors, forming simple crystalline structures, follow the so-called Matthias rule. When the superconducting critical temperature T_c is plotted as a function of valence electron count per atom (VEC), this rule shows broad peak structures at the specified VEC [15]. On the other hand, transition metal amorphous superconductors do not follow this rule and frequently show relatively high T_c values in the valley of the curve of the Matthias rule [16]. HEA superconductors with simple crystal structures have been found in bcc [17–22] and hcp [23–26]. The T_c vs. VEC plots of these HEAs seem to fall between a crystalline curve and an amorphous one [27, 28]. Thus, HEA superconductors will shed light on the study of the relationship between crystalline and amorphous compounds.

In the typical HEAs with fcc, bcc, or hcp structure, the superconductivity seems to appear in bcc or hcp HEAs. According to the classification by VEC, single-phase fcc HEA is stabilized for VEC larger than 8.0 [1, 2], where T_c would be substantially reduced in the VEC dependence of T_c observed in the Matthias rule. So it may be unrealistic to search for an fcc HEA superconductor. However, this is valuable to challenge because an fcc HEA superconductor would contribute to the deep understanding of HEA and/or the relationship between crystalline and amorphous compounds. In this chapter, we introduce our attempt at the search for an fcc HEA superconductor. Our strategy is to employ a rather high T_c element because the HEA superconductors reported to date contain superconducting elements [28]. We focused on the Nb-containing HEAs.

The concept of HEA is now used in superconducting materials with the crystal structures possessing multiple Wyckoff positions. For example, CsCl-type, α-Mn-type, A15, NaCl-type, σ-phase and $CuAl_2$-type HEA superconductors are reported

[29–37]. High degree-of-freedom in such a multisite HEA design would promote the investigations of multisite HEA superconductors. The second purpose of this chapter is the materials research on the hexagonal Mn_5Si_3-type HEAs, possessing multiple Wyckoff positions. Recently, several superconductors with the Mn_5Si_3-type—or its ordered derivative Ti_5Ga_4-type—structure have been found and attract much attention [38–44]. Besides, many intermetallic compounds are crystallizing into these crystal structures [45, 46]. **Figure 1(b)** shows the crystal structure of the Mn_5Si_3-type compound represented by M_5X_3. The space group is $P6_3/mcm$, and the M atoms occupy the 4*d* (for M1 atom) and 6*g* (for M2 atom) Wyckoff positions and the X atom another 6*g* one. The M2 atoms form a face-sharing octahedral chain along the *c*-axis. The X atoms also form another octahedron, which encloses the M1 atom forming a one-dimensional atomic chain along the *c*-axis.

In this chapter, we report the synthesis and characterization of the fcc and the Mn_5Si_3-type HEA samples. The measurement of AC magnetic susceptibility checked the superconducting state. We also present the phase analyses of both kinds of samples. Finally, the future direction of materials research on superconducting HEAs is mentioned.

2. Materials and methods

All samples were synthesized by a home-made arc furnace in an Ar atmosphere. The constituent elements as listed in **Table 1** were arc-melted on a water-cooled Cu hearth. The samples were turned over and melted several times. The Mn_5Si_3-type HEAs were annealed at 800°C for four days in evacuated quartz tubes.

A powder X-ray diffractometer (XRD-7000 L, Shimadzu, Kyoto, Japan) with Cu-Kα radiation was employed to detect the X-ray diffraction (XRD) patterns of

Element	Supply company	Purity (%)	Crystal structure	Atomic radius (Å)	VEC
Zr	Soekawa Chemicals, Tokyo, Japan	99	A3 (hcp)	1.6025	4
Nb	Nilaco, Tokyo, Japan	99.9	A2 (bcc)	1.429	5
V	Kojundo Chemical Laboratory, Sakado, Japan	99.9	A2 (bcc)	1.316	5
Ru	Soekawa Chemicals, Tokyo, Japan	99.9	A3 (hcp)	1.3384	8
Ir	Furuya Metal, Tokyo, Japan	99.99	A1 (fcc)	1.3573	9
Rh	Soekawa Chemicals, Tokyo, Japan	99.9	A1 (fcc)	1.345	9
Pd	Tanaka Kinzoku Kogyo, Tokyo, Japan	99.9	A1 (fcc)	1.3754	10
Cu	Soekawa Chemicals, Tokyo, Japan	99.99	A1 (fcc)	1.278	11
Sc	Furuya Metal, Tokyo, Japan	99.9	A3 (hcp)	1.641	3
Ti	Nilaco, Tokyo, Japan	99.9	A3 (hcp)	1.4615	4
Ga	Kojundo Chemical Laboratory, Sakado, Japan	99.99	A11	1.392	3
Si	Soekawa Chemicals, Tokyo, Japan	99.999	A4	1.153	4
Ge	Soekawa Chemicals, Tokyo, Japan	99.999	A4	1.24	4
Pt	Tanaka Kinzoku Kogyo, Tokyo, Japan	99.9	A1 (fcc)	1.387	10

Table 1.
Materials used in this study. The supply company, purity, crystal structure at room temperature, atomic radius [47], and VEC are also listed.

prepared samples. The microstructure of each sample was examined by a field emission scanning electron microscope (FE-SEM, JSM-7100F; JEOL, Akishima, Japan). The atomic compositions of the samples were checked by an energy dispersive X-ray (EDX) spectrometer equipped to the FE-SEM.

To confirm the diamagnetic signal due to the superconducting state, the temperature dependence of the AC magnetic susceptibility χ_{ac} (T) was measured by a home-made system in a GM refrigerator (UW404, Ulvac cryogenics, Kyoto, Japan) between 3 and 300 K. The amplitude and the frequency of the AC field were 5 Oe and 800 Hz, respectively.

3. Results and discussion

3.1 Nb-containing fcc HEAs

The starting compositions of prepared Nb-containing samples were determined, considering the conventional design rule [1, 2]: a δ-value less than 5% and a VEC larger than 8.0. To realize the requirements, fcc elements were predominantly used (see also **Tables 1** and **2**). The parameters of δ and VEC were calculated as follows:

No.	Sample	Composition of Phase I, II or III	δ	VEC
1	$Cu_{20}Nb_{15}Pd_{25}Rh_{30}V_{10}$		3.52	8.65
	Phase I	$Cu_{8.3(8)}Nb_{21.2(8)}Pd_{21.6(4)}Rh_{42.9(5)}V_{6.0(5)}$	3.14	8.29
	Phase II	$Cu_{14.4(5)}Nb_{12.6(5)}Pd_{28(1)}Rh_{26.4.(8)}V_{18.6(5)}$	3.26	8.32
	Phase III	$Cu_{65(5)}Nb_{2(1)}Pd_{28(2)}Rh_{2(1)}V_{3(1)}$	3.56	10.38
2	$Cu_{21}Ir_{21}Nb_{15}Pd_{22}Rh_{21}$		3.45	9.04
	Phase I	$Cu_{1.4(6)}Ir_{36.4(8)}Nb_{24.3(4)}Pd_{10.1(9)}Rh_{27.8(3)}$	2.52	8.16
	Phase II	$Cu_{8.5(7)}Ir_{6(1)}Nb_{12.5(6)}Pd_{46(3)}Rh_{27(2)}$	2.66	9.13
	Phase III	$Cu_{40(5)}Nb_{4(1)}Pd_{52(3)}Rh_{4(1)}$	3.74	10.16
3	$Cu_{21}Nb_{15}Pd_{22}Rh_{21}Zr_{21}$		8.00	7.99
	Phase I	$Cu_{19.3(3)}Pd_{37.2(2)}Rh_{19.2(4)}Zr_{24.3(5)}$	8.30	8.54
	Phase II	$Cu_{7(1)}Nb_{41(1)}Pd_{8(1)}Rh_{27(1)}Zr_{17(1)}$	6.60	6.73
	Phase III	$Cu_{57(1)}Pd_{13(1)}Rh_{9(1)}Zr_{21(1)}$	9.33	9.22
4	$Cu_{20}Nb_{15}Pd_{24}Rh_{25}V_{10}Zr_{6}$		5.62	8.34
	Phase I	$Cu_{14.5(3)}Nb_{15.7(5)}Pd_{31(1)}Rh_{23.1(5)}V_{7(1)}Zr_{8.7(5)}$	5.98	8.26
	Phase II	$Cu_{15.0(5)}Nb_{24(1)}Pd_{18(1)}Rh_{27.2(5)}V_{15.8(8)}$	3.72	7.89
	Phase III	$Cu_{86(1)}Pd_{14(1)}$	2.62	10.86
5	$Cu_{40}Nb_{20}Pd_{30}V_{10}$		4.44	8.9
	Phase I	$Cu_{21.1(2)}Nb_{27.6(2)}Pd_{39.7(7)}V_{11\ 5(7)}$	4.05	8.25
	Phase II	$Cu_{89.2(5)}Pd_{10.8(5)}$	2.35	10.89
6	$Ir_{10}Nb_{17}Pd_{33}Rh_{28}Ru_{12}$		2.21	8.54
	Phase I	$Ir_{15.5(2)}Nb_{17.7(8)}Pd_{20.3(7)}Rh_{29.7(2)}Ru_{16.8(5)}$	2.30	8.33
	Phase II	$Ir_{4.0(6)}Nb_{16.0(5)}Pd_{51(1)}Rh_{22(1)}Ru_{7.0(7)}$	2.02	8.8

Table 2.
δ and VEC of Nb-containing samples and phases detected by EDX measurements.

$$\delta = 100 \times \sqrt{\sum_{i=1}^{n} c_i \left(1 - \frac{r_i}{\bar{r}}\right)^2} \tag{1}$$

and

$$VEC = \sum_{i=1}^{n} c_i VEC_i, \tag{2}$$

where c_i, r_i, and VEC_i are the atomic fraction, the atomic radius, and the VEC of element i, respectively, and $\bar{r}$ is the composition-weighted average atomic radius. The parameter δ means the degree of the atomic size difference among the constituent elements. The calculated parameters for the prepared samples are listed in **Table 2**, in which the samples are named as their starting compositions. $Cu_{20}Nb_{15}Pd_{25}Rh_{30}V_{10}$, $Cu_{21}Ir_{21}Nb_{15}Pd_{22}Rh_{21}$ and $Ir_{10}Nb_{17}Pd_{33}Rh_{28}Ru_{12}$ fulfill the design requirements. To investigate the effect of larger δ values, two samples, including a Zr atom, were prepared.

Shown in **Figure 2** is the XRD patterns of prepared samples. In the upper five samples, all containing Nb, Pd, and Cu atoms, $Cu_{20}Nb_{15}Pd_{25}Rh_{30}V_{10}$ and $Cu_{21}Ir_{21}Nb_{15}Pd_{22}Rh_{21}$ possess dominant fcc phases. On the other hand, the XRD patterns of Zr-containing samples ($Cu_{21}Nb_{15}Pd_{22}Rh_{21}Zr_{21}$ and $Cu_{20}Nb_{15}Pd_{24}Rh_{25}V_{10}Zr_6$) cannot be characterized by fcc phases. These results suggest that Zr is unfavorable for the formation of an fcc structure. In order to further investigate the formation condition of the single fcc phase, the quaternary alloy $Cu_{40}Nb_{20}Pd_{30}V_{10}$ was synthesized. As shown in **Figure 2**, this sample exhibits two fcc phases with quite different lattice parameters. The XRD pattern of the sample with no Cu atom

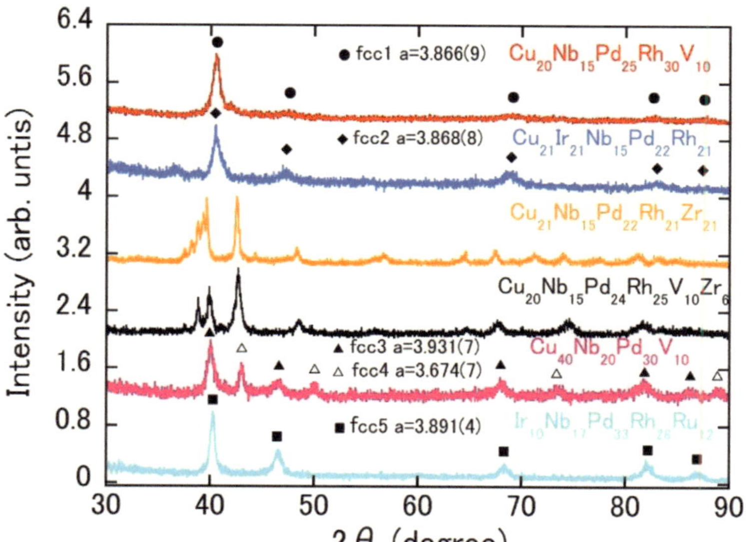

Figure 2.
XRD patterns of Nb-containing samples. The origin of each pattern is shifted by an integer value.

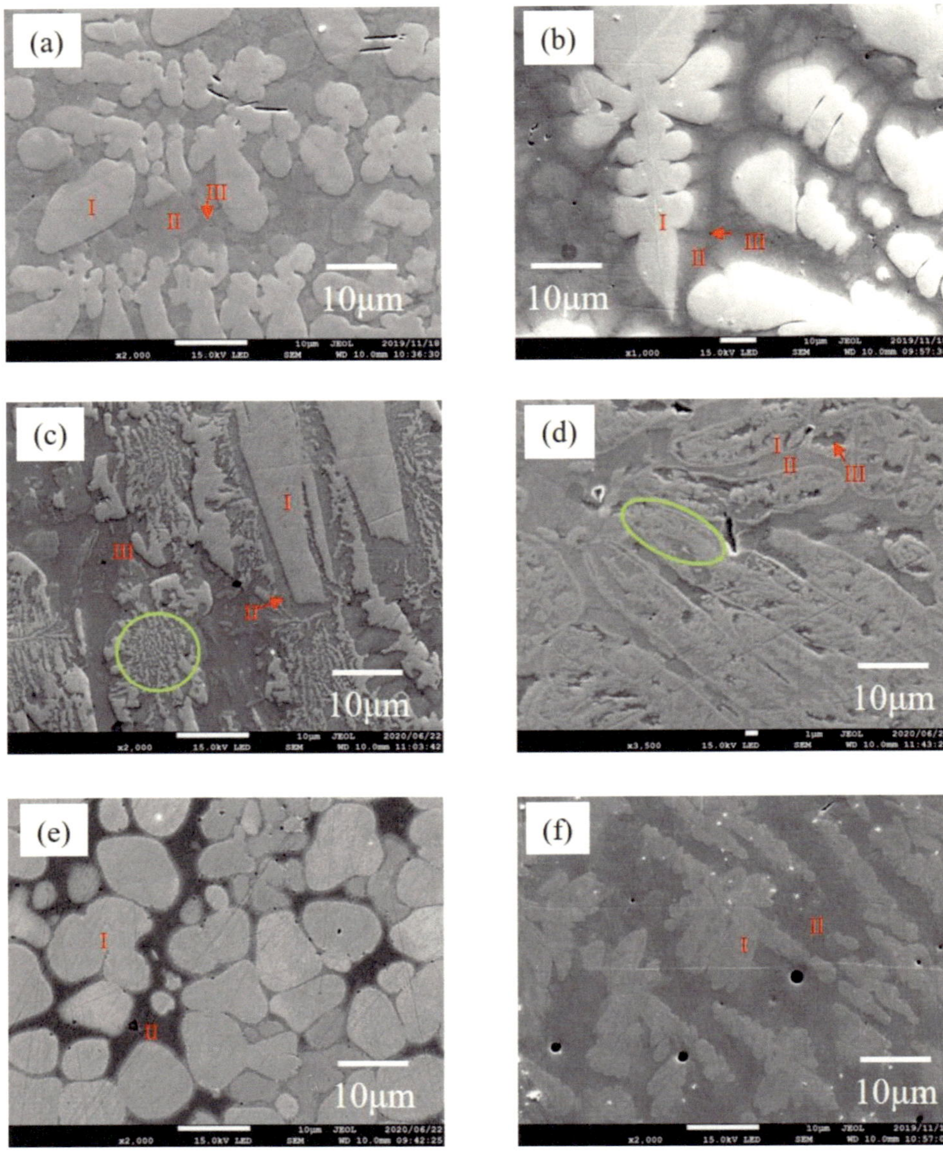

Figure 3.
Back-scattered electron (15 keV) images of (a) $Cu_{20}Nb_{15}Pd_{25}Rh_{30}V_{10}$*, (b)* $Cu_{21}Ir_{21}Nb_{15}Pd_{22}Rh_{21}$*, (c)* $Cu_{21}Nb_{15}Pd_{22}Rh_{21}Zr_{21}$*, (d)* $Cu_{20}Nb_{15}Pd_{24}Rh_{25}V_{10}Zr_{6}$*, (e)* $Cu_{40}Nb_{20}Pd_{30}V_{10}$*, and (f)* $Ir_{10}Nb_{17}Pd_{33}Rh_{28}Ru_{12}$*, respectively.*

(see the bottom of **Figure 2**) can be explained by an fcc phase. The lattice parameters of all fcc phases were obtained by the least-square method [48, 49] and are shown in **Figure 2**.

Figures 3(a)–(f) display the SEM images of samples, all indicating multi-phases. In $Cu_{20}Nb_{15}Pd_{25}Rh_{30}V_{10}$ (**Figure 3(a)**) and $Cu_{21}Ir_{21}Nb_{15}Pd_{22}Rh_{21}$ (**Figure 3(b)**), three contrast phases I, II and III were detected. In each case, the brightest area (phase I) showed a dendritic morphology, which is surrounded by phase II with the median contrast. The darkest area (phase III) would be the precipitate that formed in the final solidification process. A part of $Cu_{21}Nb_{15}Pd_{22}Rh_{21}Zr_{21}$ (**Figure 3(c)**) or $Cu_{20}Nb_{15}Pd_{24}Rh_{25}V_{10}Zr_{6}$ (**Figure 3(d)**) showed a eutectic-like structure formed by phase I and phase II (see, for example, the green elliptic closed-curve). As shown in **Figure 3(e)**, $Cu_{40}Nb_{20}Pd_{30}V_{10}$ possesses two phases, both of which would be fcc phases taking into account the XRD results. $Ir_{10}Nb_{17}Pd_{33}Rh_{28}Ru_{12}$ displays two contrast areas (see phases I and II in **Figure 3(f)**). The shape of the main phase has a dendritic-like morphology. The compositions of all phases determined by EDX are listed in **Table 2**.

χ_{ac} (T) measurements of all prepared samples suggested no superconductivity down to 3 K, which means that a Nb-containing fcc-HEA might be an inadequate strategy for searching fcc HEA superconductors. The appearance of superconductivity in a Nb-containing fcc compound might be a rare event because almost all Nb-based superconductors form bcc-related structures. NbN or NbC superconductor is a rare example, crystallizing into a NaCl-type structure related to the fcc structure [50, 51]. Although our results would be negative for the research of Nb-containing fcc HEA superconductors, it is to be noted that phase II in sample no. 1, phase I in sample no. 6 and possibly phases I and II in sample no. 4 are new members of HEA.

Here, we discuss the fcc phase stability, viewed from the parameters of δ and VEC, which is summarized in **Table 2**. We also calculate these parameters for the phases detected by EDX. The values of δ were very large in $Cu_{21}Nb_{15}Pd_{22}Rh_{21}Zr_{21}$ and $Cu_{20}Nb_{15}Pd_{24}Rh_{25}V_{10}Zr_6$ due to the larger atomic radius of Zr, which would lead to no fcc phase in each sample. While δ is reduced in $Cu_{20}Nb_{15}Pd_{25}Rh_{30}V_{10}$ and $Cu_{21}Ir_{21}Nb_{15}Pd_{22}Rh_{21}$, each sample shows three phases, all with rather low δ values. In each case, going from phase I to phase III, the VEC value increases, which accompanies the decrease (increase) of the Nb (Cu) atomic fraction. This suggests that the combination of Nb and Cu is not recommended even with a reduced δ, because an Nb-rich phase and a Cu-rich phase are stabilized for a smaller VEC and a larger VEC, respectively. Probably due to this reason, quaternary $Cu_{40}Nb_{20}Pd_{30}V_{10}$ does not show a single fcc phase. We note here that the addition of Cu leads to the breakdown of single-phase fcc CoCrFeNi into two fcc phases. This is ascribed to the positive enthalpy of mixing between the Cu and several elements [52]. $Ir_{10}Nb_{17}Pd_{33}Rh_{28}Ru_{12}$ with substantially suppressed δ and no Cu atom was expected to show a single fcc phase; however, two phases were detected in the sample. The detected phases possess reduced δ values and a similar VEC. Thus, it may not be easy to synthesize a single-phase Nb-containing fcc HEA.

No.	Sample	Composition of main phase	Composition of minor phase	VEC
1	(NbScTiVZr)(GaGeSi)	$(Nb_{13.0(1)}SC_{15.5(1)}Ti_{11.2(1)}V_{4.6(2)}Zr_{19.0(1)})(Ga_{4.4(2)}Ge_{19.3(1)}Si_{13.1(1)})$	$Ga_{7(1)}Ge_{6(1)}Nb_{19(1)}Sc_{8(1)}Si_{7(1)}Ti_{17(1)}V_{28(1)}Zr_{8(1)}$	4
2	$(Nb_{1.25}Sc_{1.25}Ti_{1.25}Zr_{1.25})(Ge_{1.8}Si_{1.2})$	$(Nb_{17(1)}Sc_{14(1)}Ti_{16(1)}Zr_{17(1)})(Ge_{22(1)}Si_{14(1)})$	—	4
3	$(Nb_{1.25}Sc_{1.25}Ti_{1.25}Zr_{1.25})(Ge_{1.55}Ru_{0.47}Si_{0.98})$	$(Nb_{18.0(2)}Sc_{13.5(4)}Ti_{12.3(5)}Zr_{20.5(6)})(Ge_{21.4(2)}Si_{14.3(4)})$	$Nb_{28(4)}Ru_{31(4)}Sc_{6(2)}T_{31(3)}Zr_{4(1)}$	4.234
4	$(Nb_{1.4}ScTiZr_{1.6})(Ge_{1.6}Pt_{0.3}Si_{1.1})$	$(Nb_{20.8(3)}Sc_{8.4(5)}Ti_{10.9(3)}Zr_{24(1)})(Ge_{19.9(5)}Pt_{1.0(4)}Si_{15.0(5)})$	$Ge_{5.5(7)}Sc_{37(2)}Si_{1.7(2)}Ti_{15(1)}Pt_{40.7(7)}$	4.275
5	$(Nb_{1.4}ScTiZr_{1.6})(Ge_{1.6}Ir_{0.3}Si_{1.1})$	$(Nb_{20.4(2)}Sc_{8.9(5)}Ti_{10.9(1)}Zr_{24.5(4)})(Ge_{20.3(3)}Si_{15.0(2)})$	$Ge_{5.3(5)}Ir_{27.7(9)}Nb_{9.3(5)}Sc_{20.5(9)}Si_{4.0(6)}Ti_{16.4(5)}Zr_{16.7(8)}$	4.238

Table 3.
Chemical compositions of phases detected by EDX measurements for Mn_5Si_3-type HEA samples. VEC of each starting composition is also shown.

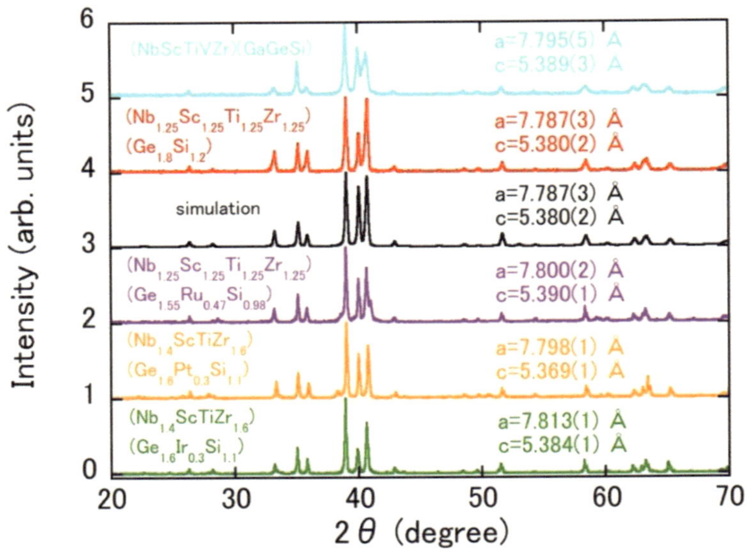

Figure 4.
XRD patterns of Mn_5Si_3-type HEAs. The simulation pattern is also shown. Each pattern is shifted by an integer value for clarity.

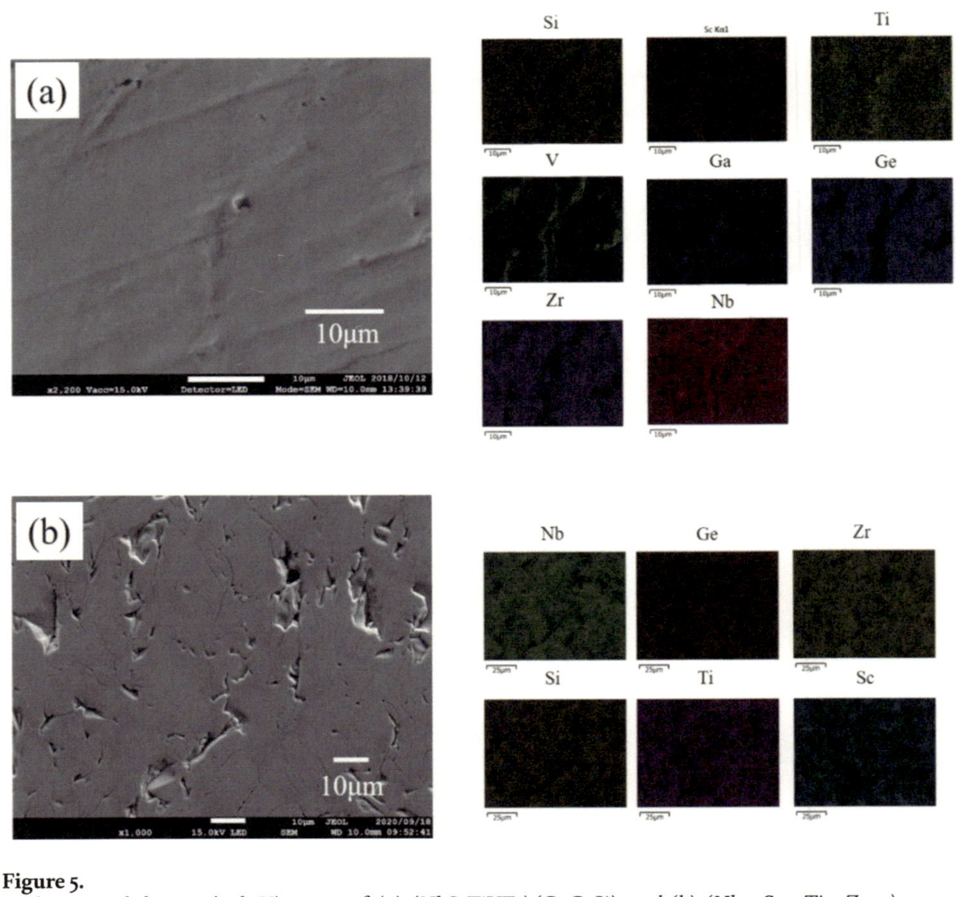

Figure 5.
Back-scattered electron (15 keV) images of (a) (NbScTiVZr)(GaGeSi), and (b) $(Nb_{1.25}Sc_{1.25}Ti_{1.25}Zr_{1.25})(Ge_{1.8}Si_{1.2})$, respectively. The elemental mappings are also shown.

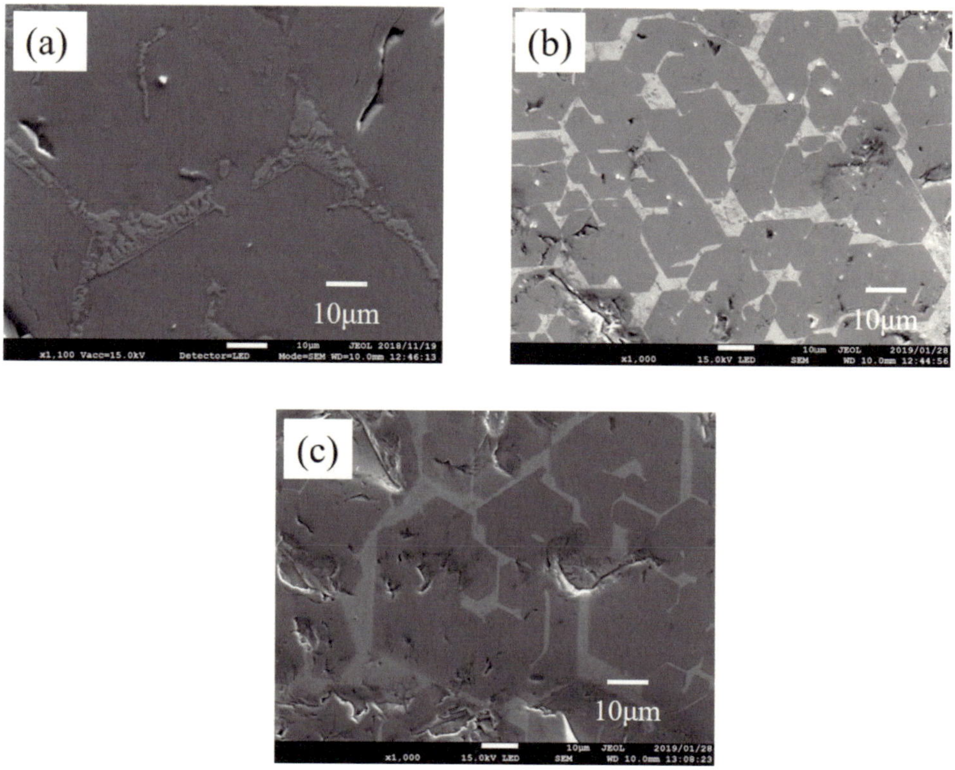

Figure 6.
Back-scattered electron (15 keV) images of (a) $(Nb_{1.25}Sc_{1.25}Ti_{1.25}Zr_{1.25})(Ge_{1.55}Ru_{0.47}Si_{0.98})$, (b) $(Nb_{1.4}ScTiZr_{1.6})(Ge_{1.6}Pt_{0.3}Si_{1.1})$, and (c) $(Nb_{1.4}ScTiZr_{1.6})(Ge_{1.6}Ir_{0.3}Si_{1.1})$, respectively.

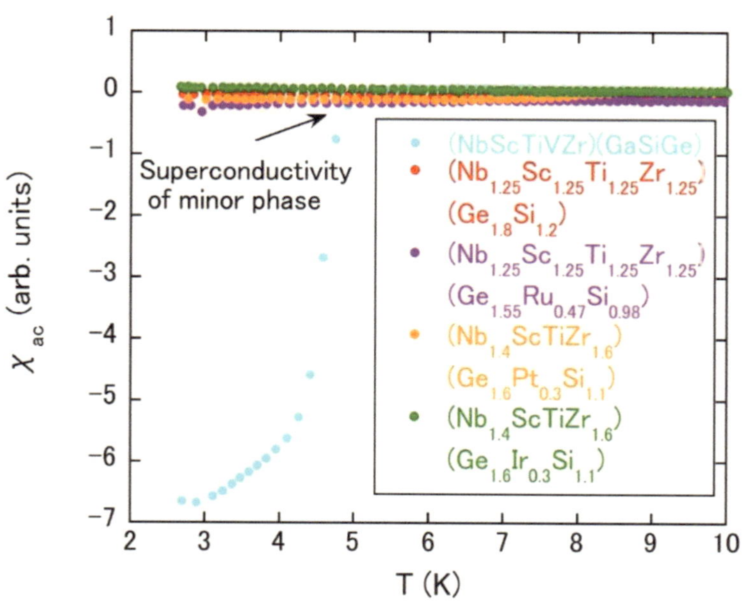

Figure 7.
Temperature dependences of χ_{ac} of Mn_5Si_3-type HEA samples.

3.2 Mn_5Si_3-type HEAs

We have prepared five Mn_5Si_3-type HEAs as listed in **Table 3**, and the XRD patterns are given in **Figure 4**. All XRD patterns are well indexed by the hexagonal Mn_5Si_3-type structure, and the determined lattice parameters are displayed in **Figure 4**. The SEM images of all samples are presented in **Figures 5** and **6**, and χ_{ac} (T) of each sample is shown in **Figure 7**.

We have started from (NbScTiVZr)(GaGeSi), which shows a diamagnetic signal (see **Figure 7**). However, as shown in **Figure 5(a)**, the elemental mapping has revealed the inhomogeneous distribution of constituent elements, which is obviously signaled by the V atom. The atomic compositions determined by EDX are $(Nb_{13.0(1)}Sc_{15.5(1)}Ti_{11.2(1)}V_{4.6(2)}Zr_{19.0(1)})(Ga_{4.4(2)}Ge_{19.3(1)}Si_{13.1(1)})$ for the V-poor phase and $Ga_{7(1)}Ge_{6(1)}Nb_{19(1)}Sc_{8(1)}Si_{7(1)}Ti_{17(1)}V_{28(1)}Zr_{8(1)}$ for the V-rich phase, respectively. The separately synthesized latter phase crystallizes into a bcc structure. This compound also shows the diamagnetic signal at approximately 5 K, which is identical to that of (NbScTiVZr)(GaGeSi). Therefore, (NbScTiVZr)(GaGeSi) would be an intrinsically normal state down to 3 K. The result of the chemical composition of the Mn_5Si_3-type phase in (NbScTiVZr)(GaGeSi) suggests the difficulty of incorporation of V and Ga atoms in a Mn_5Si_3-type HEA. Taking into account this experimental result, we have synthesized $(Nb_{1.25}Sc_{1.25}Ti_{1.25}Zr_{1.25})(Ge_{1.8}Si_{1.2})$. As shown in **Figure 4**, the sample is almost single phase, which is also supported by homogeneous elemental mapping (see also **Figure 5(b)**). The determined atomic composition is $(Nb_{17(1)}Sc_{14(1)}Ti_{16(1)}Zr_{17(1)})(Ge_{22(1)}Si_{14(1)})$, which agrees well with the starting composition. While the single-phase Mn_5Si_3-type HEA is successfully obtained, the diamagnetic signal cannot be confirmed down to 3 K, as shown in **Figure 7**.

One of the conceivable reasons for no superconductivity in the samples mentioned above is that the VEC value is slightly less than the optimal value (see also **Table 3**). As pointed out in the review [28], multisite HEA superconductors follow the respective Matthias rule, which means the important role of the density of states at the Fermi level. The VEC values of Mn_5Si_3-type superconductors Zr_5Sb_3 and $Zr_5Ge_{2.5}Ru_{0.5}$ are 4.375 and 4.25, respectively [40, 42], while the VEC value of (NbScTiVZr)(GaGeSi) or $(Nb_{1.25}Sc_{1.25}Ti_{1.25}Zr_{1.25})(Ge_{1.8}Si_{1.2})$ is 4. Thus, aiming at increasing the VEC, we substituted Ru, Pt, or Ir atoms at the Si site of Mn_5Si_3-type HEA. The prepared samples were $(Nb_{1.25}Sc_{1.25}Ti_{1.25}Zr_{1.25})(Ge_{1.55}Ru_{0.47}Si_{0.98})$, $(Nb_{1.4}ScTiZr_{1.6})(Ge_{1.6}Pt_{0.3}Si_{1.1})$, and $(Nb_{1.4}ScTiZr_{1.6})(Ge_{1.6}Ir_{0.3}Si_{1.1})$ with the respective VEC value of 4.234, 4.275, and 4.238. In each sample, the main phase of XRD pattern is well characterized by the Mn_5Si_3-type structure (see **Figure 4**). However, the atomic composition, deviating from the starting one, as shown in **Table 3**, indicates that Ru, Pt, or Ir atoms cannot replace the atoms at the Si site. The SEM images of these samples show the precipitation of impurity phases at the grain boundaries of hexagonal-shaped main phases (see **Figure 6(a)–(c)**). χ_{ac} (T) measurements of these Mn_5Si_3-type HEA do not show a superconducting signal down to 3 K.

4. Summary

We have carried out materials research on the fcc and the Mn_5Si_3-type HEA superconductors. In the study of fcc HEA superconductors, we employed the Nb element, taking into account that the inclusion of rather high T_c elements is advantageous. Although some Nb-containing samples showed dominant fcc phases, single-phase ones could not be obtained. While we have found several new Nb-containing HEA phases in the multi-phase samples, no superconductivity appeared in each HEA phase down to 3 K. Considering that the Nb element forms a bcc structure at room temperature, and there are only several examples of fcc-related Nb-based superconductors, the discovery of Nb-containing fcc HEA superconductors would be a hard task. For the Mn_5Si_3-type HEA samples, we have successfully confirmed the single phase in $(Nb_{1.25}Sc_{1.25}Ti_{1.25}Zr_{1.25})(Ge_{1.8}Si_{1.2})$ with VEC = 4. However, this HEA compound does not show the diamagnetic signal down to 3 K. Some Mn_5Si_3-type superconductors possess the VEC value of 4.25 or 4.375. Thus Ru, Ir, or Pt element was substituted at the Si site to increase VEC, resulting in the unsuccessful attempt. We need a strategy to adjust the VEC value for the Mn_5Si_3-type HEAs.

The conclusions regarding the manufacturing process or microstructure in HEA superconductors are bulleted below.

- If one wants to obtain a single-phase sample with bcc, hcp, or fcc type structure by the arc-melting method, it would be a rather hard task due to an appearance of secondary phase and/or of phase with a slightly different composition.

- Eutectic HEAs receive much attention due to the rich functions arising from the microstructures [53]. In some cases, eutectic superconductors show enhanced superconducting critical temperatures. Therefore, the study of the eutectic phase in HEA superconductors might be interesting.

- Mechanical alloying has been widely used to produce HEAs [54]. The mechanical alloying process is different from the arc-melting one. So this is another route to obtain single-phase HEA superconductors.

5. Future directions

The formation of single-phase fcc HEA is realized at VEC larger than 8.0. According to the Matthias rule of transition metal alloys, T_c at such a large VEC value is substantially reduced. Therefore, the measurement of physical properties at much lower than 3 K is desired. Because the synthesis of quinary alloy with the single-phase might be a hard task, research on ternary or quarternary fcc multicomponent superconductor would be necessary. In the research area of HEAs, the CALPHAD (calculation of phase diagram) method is rapidly used for the prediction of HEAs or the study of the phase relation between HEAs and other alloys. If the thermodynamic data of various compounds in the present Nb-containing multicomponent systems are sufficiently collected, the CALPHAD method will elucidate the stability of a HEA in this system. Thus, our results will greatly assist in the evaluation of the CALPHAD method in the future.

Nb_5Ir_3O, crystallizing into the Ti_5Ga_4-type structure, which is the ordered derivative of Mn_5Si_3-type structure, is well known as a two-band superconductor [41]. By substituting Pt into Ir, the crossover to single-band superconductivity is observed [44], which is a rare phenomenon. This result promotes us to investigate multisite HEA superconductors for further search of the crossover phenomenon, and the high-entropy state may be a new route of controlling the superconducting band. Another interesting aspect of HEA is the cocktail effect. In bcc HEA superconductors, we have shown that the peculiar enhancement of T_c by introducing the high-entropy state is not observed [28]. It would be interesting to pursue the cocktail effect of T_c in multisite HEAs.

Acknowledgements

J.K. is grateful for the support provided by Comprehensive Research Organization of Fukuoka Institute of Technology.

Conflict of interest

The authors declare no conflict of interest.

Author details

Jiro Kitagawa*, Naoki Ishizu and Shusuke Hamamoto
Department of Electrical Engineering, Faculty of Engineering, Fukuoka Institute of Technology, Fukuoka, Japan

*Address all correspondence to: j-kitagawa@fit.ac.jp

References

[1] Gao M C, Yeh J W, Liaw P K, Zhang Y. High-entropy alloys: fundamentals and applications Swizerland: Springer; 2015.

[2] Murty B S, Yeh J W, Ranganathan S, Bhattacharjee P P. High-Entropy Alloys. 2nd ed. Amsterdam: Elsevier; 2019.

[3] Yeh J W, Chen S K, Lin S J, Gan J Y, Chin T S, Shun T T, Tsau C H, Chang S Y. Nanostructured high-entropy alloys with multiple principal elements: Novel alloy design concepts and outcomes. Adv. Eng. Mater. 2004; 6:299-303. DOI: 10.1002/adem.200300567

[4] Otto F, Yang Y, Bei H, George E P. Relative effects of enthalpy and entropy on the phase stability of equiatomic high-entropy alloys. Acta Mater. 2013;61:2628-2638. DOI: 10.1016/j.actamat.2013.01.042

[5] Rost C M, Sachet E, Borman T, Moballegh A, Dickey E C, Hou D, Jones J L, Curtarolo S, Maria J P. Entropy-stabilized oxides. Nat. Commun. 2015;6:8485. DOI: 10.1038/ncomms9485

[6] Musicó B L, Gilbert D, Ward T Z, Page K, George E, Yan J, Mandrus D, Keppens V. The emergent field of high entropy oxides: Design, prospects, challenges, and opportunities for tailoring material properties. APL Materials 2020;8:040912. DOI: 10.1063/5.0003149

[7] Ying T, Yu T, Shiah Y-S, Li C, Li J, Qi Y, Hosono H. High-entropy van der Waals materials formed from mixed metal dichalcogenides, halides, and phosphorus trisulfides. J. Am. Chem. Soc. 2021;143:7042-7049. DOI: 10.1021/jacs.1c01580

[8] Marik S, Singh D, Gonano B, Veillon F, Pelloquin D, Bréard Y. Enhanced magnetic frustration in a new high entropy diamond lattice spinel oxide. Scr. Mater. 2020;186:366-369. DOI: 10.1016/j.scriptamat.2020.04.027

[9] Sun L, Luo Y, Ren X, Gao Z, Du T, Wu Z, Wang J. A multicomponent γ-type $(Gd_{1/6}Tb_{1/6}Dy_{1/6}Tm_{1/6}Yb_{1/6}Lu_{1/6})_2Si_2O_7$ disilicate with outstanding thermal stability. Mater. Res. Lett. 2020;8:424-430. DOI: 10.1080/21663831.2020.1783007

[10] Rao Z, Dutta B, Körmann F, Lu W, Zhou X, Liu C, Kwiatkowski da Silva A, Wiedwald U, Spasova M, Farle M, Ponge D, Gault B, Neugebauer J, Raabe D, Li Z. Beyond solid solution high-entropy alloys: tailoring magnetic properties via spinodal decomposition. Adv. Funct. Mater. 2021;31:2007668. DOI: 10.1002/adfm.202007668

[11] Quintana-Nedelcos A, Leong Z, Morley N A. Study of dual-phase functionalisation of NiCoFeCr-Al_x multicomponent alloys for the enhancement of magnetic properties and magneto-caloric effect. Mater. Today: Energy 2021;20:100621. DOI: 10.1016/j.mtener.2020.100621

[12] Jung C, Kang K, Marshal A, Gokuldoss Pradeep K, Seol J–B, Lee H M, Choi P–P. Effects of phase composition and elemental partitioning on soft magnetic properties of AlFeCoCrMn high entropy alloys. Acta Mater. 2019;171:31-39. DOI: 10.1016/j.actamat.2019.04.007

[13] BaBa K, Ishizu N, Nishizaki T, Kitagawa J. Magnetic and transport properties of new dual-phase high-entropy alloy FeRhIrPdPt. Materials 2021;14:2877. DOI: 10.3390/ma14112877

[14] Koželj P, Vrtnik S, Jelen A, Jazbec S, Jagličić Z, Maiti S, Feuerbacher M, Steurer W, Dolinšek J. Discovery of a superconducting high-entropy alloy. Phys. Rev. Lett. 2014;113:107001. DOI: 10.1103/PhysRevLett.113.107001

[15] Matthias B T. Empirical relation between superconductivity and the number of valence electrons per atom. Phys Rev 1955;97:74-76. DOI: 10.1103/PhysRev.97.74

[16] Collver M M, Hammond R H. Superconductivity in "amorphous" transition-metal alloy films. Phys Rev Lett 1973;30:92-95. DOI: 10.1103/PhysRevLett.30.92

[17] von Rohr F O, Winiarski M J, Tao J, Klimczuk T, Cava R J. Effect of electron count and chemical complexity in the Ta-Nb-Hf-Zr-Ti high-entropy alloy superconductor. Proc. Natl. Acad. Sci. 2016;113:E7144-E7150. DOI: 10.1073/pnas.1615926113

[18] von Rohr F O, Cava R J. Isoelectronic substitutions and aluminium alloying in the Ta-Nb-Hf-Zr-Ti high-entropy alloy superconductor. Phys. Rev. Mater. 2018;2:034801. DOI: 10.1103/PhysRevMaterials.2.034801

[19] Marik S, Varghese M, Sajilesh K P, Singh D, Singh R P. Superconductivity in equimolar Nb-Re-Hf-Zr-Ti high entropy alloy. J. Alloys Compd. 2018;769:1059-1063. DOI: 10.1016/j.jallcom.2018.08.039

[20] Ishizu N, Kitagawa J. New high-entropy alloy superconductor $Hf_{21}Nb_{25}Ti_{15}V_{15}Zr_{24}$. Res. Phys. 2019;13:102275. DOI: 10.1016/j.rinp.2019.102275

[21] Zhang X, Winter N, Witteveen C, Moehl T, Xiao Y, Krogh F, Schilling A, von Rohr F O. Preparation and characterization of high-entropy alloy $(TaNb)_{1-x}(ZrHfTi)_x$ superconducting films. Phys. Rev. Res. 2020;2:013375. DOI: 10.1103/PhysRevResearch.2.013375

[22] Harayama Y, Kitagawa J. Superconductivity in Al-Nb-Ti-V-Zr multicomponent alloy. J. Supercond. Nov. Magn. 2021 DOI:10.1007/s10948-021-05966-z

[23] Lee Y S, Cava R J. Superconductivity in high and medium entropy alloys based on MoReRu. Physica C 2019;566:1353520. DOI: 10.1016/j.physc.2019.1353520

[24] Marik S, Motla K, Varghese M, Sajilesh K P, Singh D, Breard Y, Boullay P, Singh R P. Superconductivity in a new hexagonal high-entropy alloy. Phys. Rev. Mater. 2019;3:060602(R). DOI: 10.1103/PhysRevMaterials.3.060602

[25] Liu B, Wu J, Cui Y, Zhu Q, Xiao G, Wu S, Cao G, Ren Z. Superconductivity in hexagonal Nb-Mo-Ru-Rh-Pd high-entropy alloys. Scr. Mater. 2020;182:109-113. DOI: 10.1016/j.scriptamat.2020.03.004

[26] Liu B, Wu J, Cui Y, Zhu Q, Xiao G, Wu S, Cao G-h, Ren Z. Structural evolution and superconductivity tuned by valence electron concentration in the Nb-Mo-Re-Ru-Rh high-entropy alloys. J. Mater. Sci. Technol. 2021;85:11-17. DOI: 10.1016/j.jmst.2021.02.002

[27] Sun L, Cava R J. High-entropy alloy superconductors: Status, opportunities, and challenges. Phys. Rev. Mater. 2019;3:090301. DOI: 10.1103/PhysRevMaterials.3.090301

[28] Kitagawa J, Hamamoto S, Ishizu N. Cutting edge of high-entropy alloy superconductors from the perspective of materials research. Metals 2020;10:1078. DOI: 10.3390/met10081078

[29] Stolze K, Tao J, von Rohr F O, Kong T, Cava R J. Sc–Zr–Nb–Rh–Pd and Sc–Zr–Nb–Ta–Rh–Pd high-entropy alloy superconductors on a CsCl-type lattice. Chem. Mater. 2018;30:906-914. DOI: 10.1021/acs.chemmater.7b04578

[30] Stolze K, Cevallos F A, Kong T, Cava R J. High-entropy alloy superconductors on an α-Mn lattice. J. Mater. Chem. C

2018;6:10441-10449. DOI: 10.1039/C8TC03337D

[31] Wu J, Liu B, Cui Y, Zhu Q, Xiao G, Wang H, Wu S, Cao G, Ren Z. Polymorphism and superconductivity in the V-Nb-Mo-Al-Ga high-entropy alloys. Sci. China Mater. 2020;63:823-831. DOI: 10.1007/s40843-019-1237-5

[32] Mizuguchi Y. Superconductivity in high-entropy-alloy telluride $AgInSnPbBiTe_5$. J. Phys. Soc. Jpn. 2019;88:124708. DOI: 10.7566/JPSJ.88.124708

[33] Kasem M R, Hoshi K, Jha R, Katsuno M, Yamashita A, Goto Y, Matsuda T D, Aoki Y, Mizuguchi Y. Superconducting properties of high-entropy-alloy tellurides M-Te (M: Ag, In, Cd, Sn, Sb, Pb, Bi) with a NaCl-type structure. Appl. Phys. Express 2020;13:033001. DOI: 10.35848/1882-0786/ab7482

[34] Yamashita A, Jha R, Goto Y, Matsuda T D, Aoki Y, Mizuguchi Y. An Efficient way of increasing the total entropy of mixing in high-entropy-alloy compounds: a case of NaCl-type $(Ag,In,Pb,Bi)Te_{1-x}Se_x$ (x = 0.0, 0.25, 0.5) superconductors. Dalton Trans. 2020;49:9118-9122. DOI: 10.1039/D0DT01880E

[35] Liu B, Wu J, Cui Y, Zhu Q, Xiao G, Wu S, Cao G-h, Ren Z. Superconductivity and paramagnetism in Cr-containing tetragonal high-entropy alloys. J. Alloys Compd. 2021;869:159293. DOI: 10.1016/j.jallcom.2021.159293

[36] Mizuguchi Y, Kasem M R, Matsuda T D. Superconductivity in $CuAl_2$-type $Co_{0.2}Ni_{0.1}Cu_{0.1}Rh_{0.3}Ir_{0.3}Zr_2$ with a high-entropy-alloy transition metal site. Mater. Res. Lett. 2020:9;141-147. DOI: 10.1080/21663831.2020.1860147

[37] Kasem M R, Yamashita A, Goto Y, Matsuda T D, Mizuguchi Y. Synthesis of high-entropy-alloy-type superconductors $(Fe,Co,Ni,Rh,Ir)Zr_2$ with tunable transition temperature. J. Mater. Sci. 2021;56:9499-9505. DOI: 10.1007/s10853-021-05921-2

[38] Cort B, Giorgi A L, Stewart G R. Low temperature specific heats of H(NbIrO) and R(NbPtO). J. Low. Temp. Phys. 1982;47:179-185. DOI: 10.1007/BF00682027

[39] Waterstrat R M, Kuentzler R, Muller J. Structural instabilities and superconductivity in quasi-binary Mn_5Si_3-type compounds. J. Less Common Met. 1990;167:169-178. DOI: 10.1016/0022-5088(90)90302-Z

[40] Lv B, Zhu X Y, Lorenz B, Wei F Y, Xue Y Y, Yin Z P, Kotliar G, Chu C W. Superconductivity in the Mn_5Si_3-type Zr_5Sb_3 system. Phys. Rev. B 2013;88:134520. DOI: 10.1103/PhysRevB.88.134520

[41] Zhang Y, Wang B, Xiao Z, Lu Y, Kamiya T, Uwatoko Y, Kageyama H, Hosono H. Electride and superconductivity behaviors in Mn_5Si_3-type intermetallics. npj Quantum Materials 2017;2:45. DOI: 10.1038/s41535-017-0053-4

[42] Li S, Liu X, Anand V, Lv B. Superconductivity from site-selective Ru doping studies in Zr_5Ge_3 compound. New J. Phys. 2018;20:013009. DOI: 10.1088/1367-2630/aa9ccd

[43] Hamamoto S, Kitagawa J. Superconductivity in oxygen-added Zr_5Pt_3. Mater. Res. Express 2018;5:106001. DOI: 10.1088/2053-1591/aad9cf

[44] Xu Y, Jöhr S, Das L, Kitagawa J, Medarde M, Shiroka T, Chang J, Shang T. Crossover from multiple- to single-gap superconductivity in $Nb_5Ir_{3-x}Pt_xO$ alloys. Phys. Rev. B 2020;101:134513. DOI: 10.1103/PhysRevB.101.134513

[45] Corbett J D, Garcia E, Guloy A M, Hurng W M, Kwon Y U, Leon-Escamilla E A. Widespread interstitial chemistry of Mn_5Si_3-type and related phases. Hidden impurities and opportunities. Chem. Mater. 1998;10:2824-2836. DOI: 10.1021/cm980223c

[46] Kitagawa J, Hamamoto S. Superconductivity in $Nb_5Ir_{3-x}Pt_xO$. JPS Conf. Proc. 2020;30:011055. DOI: 10.7566/JPSCP.30.011055

[47] Miracle D B, Senkov O N. A critical review of high entropy alloys and related concepts. Acta Mater. 2017;122:448-511. DOI: 10.1016/j.actamat.2016.08.081

[48] Izumi F, Momma K. Three-dimensional visualization in powder diffraction. Solid State Phenom. 2007; 130:15-20. DOI: 10.4028/www.scientific.net/SSP.130.15

[49] Tsubota M, Kitagawa J. A necessary criterion for obtaining accurate lattice parameters by Rietveld method. Sci. Rep. 2017;7:15381. DOI: 10.1038/s41598-017-15766-y

[50] Giorgi A L, Szklarz E G, Storms E K, Bowman A L, Matthias B T. Effect of Composition on the superconducting transition temperature of tantalum carbide and niobium carbide. Phys. Rev. 1962;125:837-838. DOI: 10.1103/PhysRev.125.837

[51] Matthias B T. Transition temperatures of superconductors. Phys. Rev. 1953;92:874-876. DOI: 10.1103/PhysRev.92.874

[52] Otto F, Yang Y, Bei H, George E P. Relative effects of enthalpy and entropy on the phase stability of equiatomic high-entropy alloys. Acta Materialia 2013;61:2628-2638. DOI: 10.1016/j.actamat.2013.01.042

[53] Lu Y, Dong Y, Guo S, Jiang L, Kang H, Wang T, Wen B, Wang Z, Jie J, Cao Z, Ruan H, Li T. A promising new class of high-temperature alloys: Eutectic high-entropy alloys. Sci. Rep. 2015;4:6200. DOI: 10.1038/srep06200

[54] Vaidya M, Muralikrishna G M, Murty B S. High-entropy alloys by mechanical alloying: A review. J. Mater. Res. 2019;34:664-686. DOI: 10.1557/jmr.2019.37

Section 2

High-Entropy Alloy Composites

Chapter 3

Why Al-B_4C Metal Matrix Composites? A Review

Mohamed F. Ibrahim, Hany R. Ammar, Agnes M. Samuel, Mahmoud S. Soliman, Victor Songmene and Fawzy H. Samuel

Abstract

The Al-B_4C metal matrix composite (MMC) is characterized by its ability to absorb neutrons which makes it the most suitable shielding material for nuclear reactors. The present work was performed on two series of Al-B_4C metal matrix composites made using a powder injection apparatus. In one series, commercially pure aluminum (A5) served as the matrix. For the second set, 6063 alloy was used. In all cases the volume fraction of B_4C reinforcement particles (grit size 400 mesh, purity 99.5%) was approximately 15%. The volume fraction of the injected B_4C particles was determined using a computer driven image analyzer. Measured amounts of Ti, Zr, and Ti + Zr, were added to the molten composites of both series. Microstructural characterization was carried out employing a field emission scanning electron microscope operating at 20 kV and equipped with an electron dispersive x-ray spectroscopic system (EDS). The same technique was applied to characterize the fracture behavior of the tested composites. Mechanical properties of these composites were investigated using impact testing, and ambient and high temperature tensile testing methods. Almost 1000 impact and tensile samples were tested following different heat treatments. The obtained results from these investigations are reported in this Chapter.

Keywords: MMC, precipitation hardening, FESEM, tensile testing, impact testing, microstructural characterization

1. Introduction

The Al-B_4C metal matrix composite (MMC) is characterized by its high thermal conductivity and its ability to absorb neutrons which makes it a suitable shielding material [1]. Increasing the concentration of B_4C (>30%) increases the composite strength as well as its neutron absorption capacity. Roy et al. [2] suggested the use of 7xxx alloys as base material for the MMC due to its low density and its hardening ability caused by heat treatment which would contribute to the strength of the MMC. The use of 2124 Al alloy composites reinforced with B_4C particulates has been proposed by Öksüz and Oskay [3]. The authors claim that the volumetric wear rates of the 2124 Al alloy and its composites are increased with increase in the applied load. Singla et al. [4] proposed the use of molten technique for the production of Al-B_4C MMC. The authors studied an MMC made of Al-7075 alloy as the matrix and B_4C 32 μm particulate as the reinforcement agent. Mohan

IntechOpen

and Kennedy [5] investigated the machinability of Al-(7 and 14) wt.% Si alloys reinforced with B_4C. The MMCs were developed using the stir casting technique. Their results show that the composite reinforced with B_4C with a particle size of 100 nanometers has better mechanical properties and wear behavior compared to those reinforced with 24-micron or 6-micron sized particulates. Vaidya et al. [6] found that the strength of B_4C particle reinforced Al 6061 composite was significantly greater than the unreinforced alloy.

Drilling experiments were conducted by Kumar et al. [7] on 6061 alloy-15%B_4C (220 μm particulate diameter) using a vertical machine with High Speed Steel drills of 6 mm, 9 mm and 12 mm diameter under dry drilling conditions. It was found that speed, design of the experiment and drill diameter have a marked influence on the Over-Cut (half the difference of the diameter of the hole produced to the tool diameter). Topcu [8] and Manjunatha et al. [9] used the powder atomization technique to produce Al-5% B_4C and Al-15%B_4C MMCs. The authors reported that the wear resistance increased in proportion to the amount of the boron carbide reinforced. Tribo-surface characteristics of two aluminum metal matrix composites (Al-MMC) of compositions Al–13 vol%B_4C and Al–13 vol%SiC sliding against a commercial phenolic brake pad under dry conditions were investigated by Shorowordi et al. [10–12]. The friction coefficient was found to decrease slightly at high contact pressure.

The wear rate and friction coefficient of Al–B_4C was lower than that of Al–SiC. Several studies on friction behavior involving Al-MMC friction against ferrous materials revealed that during sliding, a layer, termed as mechanically mixed layer (MML), was formed on the worn surface of the Al-MMC [13–17]. Such layer, however, was not found to form on unreinforced aluminum. Several researchers [18–22] studied the production of Al-11%B_4C using stir melt technique. The 6061 alloy was the matrix to which B_4C particles were added. Prior to addition, the B_4C particles were preheated along with K_2TiF_6 halide salt. The resulting composite was found to have improved mechanical properties compared to the base alloy. Uthayakumar et al. [23] performed a study on the wear performance of Al–5%SiC–5%B_4C hybrid composites under dry sliding conditions using a pin on disc tribometer method. The main conclusion was that the hybrid composites can retain the wear resistance properties up to 60 N load and sliding speed ranges of 1–4 m/s.

Comparison of microstructural and mechanical properties of Al–10 vol% TiC, Al–10 vol% B_4C and Al–5 vol% TiC–5 vol% B_4C composites prepared by casting techniques was made by Mazaheri et al. [24]. The results show that the wear behavior of Al-B_4C MMC is the best among the three composites studied. The wettability of B_4C particulates was investigated by Toptan et al. [25]. They found that addition of Ti leads to formation of thin layers (80–180 nm in thickness) of Ti-C and Ti-B around the B_4C particulates which would solve the wettability issue. Similar observation on titanium as one of the reactive metals that can be used to increase wettability in Al-B_4C system was reported by other researchers [26–31]. According to Wang et al. [32] and Yang et al. [33], the stress distribution within a particle-reinforced composite subjected to external loading is non-uniform. Nanostructured Al–B_4C composite sheets were processed by accumulative roll bonding (ARB), and the effect of the number of ARB cycles on the distribution of the B_4C particles in the Al matrix was evaluated by Yazdani and Salahinejad [34] who noted an improvement in the reinforcement distribution by increasing the ARB cycles.

The present chapter summarizes the work that was carried out by the present authors using two types of Al-B_4C composites: (i) a mechanically alloyed composite supplied by Ceradyne Canada ULC, a 3 M Company, Chicoutimi, Québec, Canada, and (ii) an in-house made composite using powder injection at the Université du Québec a Chicoutimi [35–44].

2. Experimental procedure

2.1 Composite preparation

Reinforcement powder (grit size 400 and 95.4% purity) additions of 15 vol. % were made using a powder injection apparatus (**Figure 1**). The B_4C particulate was injected into molten Al. Fe, Ti and Zr additions were introduced into the molten bath, using Al-25%Fe, Al-10% Ti and Al-15% Zr master alloys, respectively, whereas Mg and Si were added as pure elements. Chemical compositions of the investigated composites are listed in **Table 1**. A general view of the powder injection set-up showing a schematic of the injection system is shown in **Figure 1**. It consists of the following components:

i. a fluidizer tube

ii. a carrier tube and a quartz nozzle

iii. resistance heating coils

iv. an adjustable two-dimensional movable stand

v. a melting unit with resistance heating

vi. an impeller (stirrer) with adjustable rotation speed

vii. flow diversion baffles.

In order to ensure uniform distribution of the B_4C particulates, the molten composite melt was stirred vigorously (300 rpm) at 730 ± 5°C. Thereafter, the molten composite was poured in two different metallic molds preheated at 450°C, as shown in **Figure 2**: an L-shaped mold (3.5x 3.8 x 30.5 cm) which was used for microstructure characterization, and a book-type mold (4 x 17 x 34 cm). In order to determine

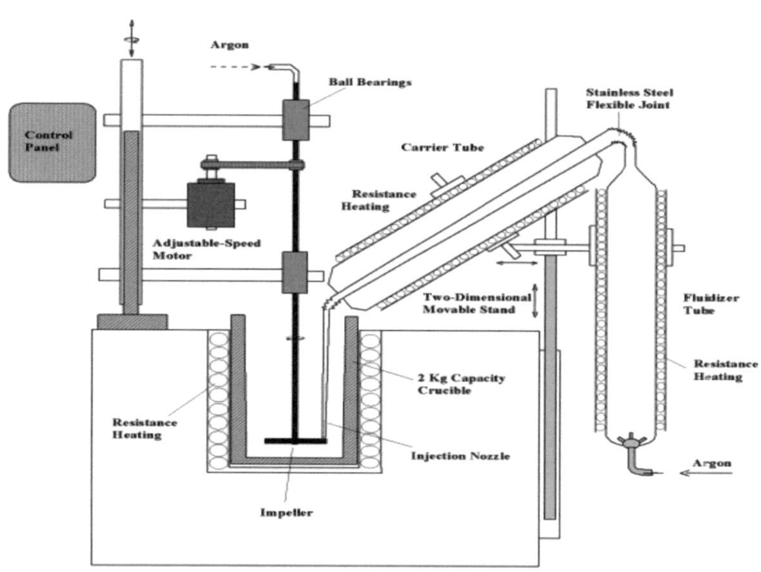

Figure 1.
A schematic diagram of the powder injector used in the present work.

Alloy Code	Composition (Ti, Zr, Sc in wt%)
1A	Al-15v/oB4C
2A	Al-15v/oB4C + 0.45%Ti
3A	Al-15v/oB4C + 0.45%Ti + 0.25%Zr
4A	Al-15v/oB4C + 0.45%Ti + 0.15%Sc
5A	Al-15v/oB4C + 0.45%Ti + 0.15%Sc + 0.25%Zr
1B	6063-15v/oB4C
2B	6063-15v/oB4C + 0.45%Ti
3B	6063-15v/oB4C + 0.45%Ti + 0.25%Zr
4B	6063-15v/oB4C + 0.45%Ti + 0.15%Sc
5B	6063-15v/oB4C + 0.45%Ti + 0.15%Sc + 0.25%Zr

Table 1.
Codes and compositions of the MMCs used in this study.

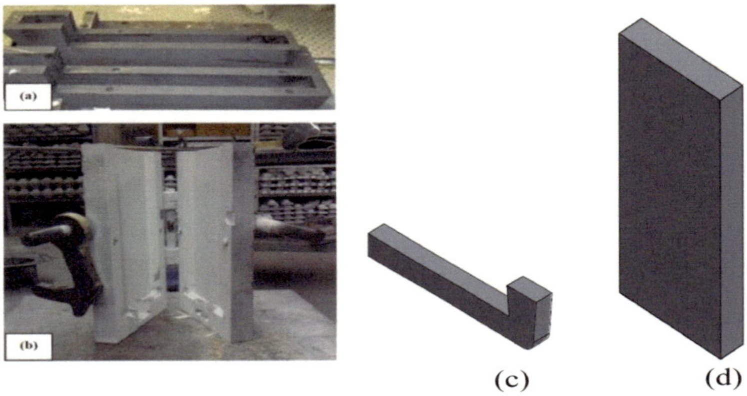

Figure 2.
(a) L-shaped mold, (b) book-type mold, (c) L-shaped casting, (d) book-mold casting.

the solidification rate obtained from each mold, trials were made using Al-7%Si. **Figure 3** depicts the dendrite arm spacing (DAS) and grain size corresponding to each mold. The castings made using the book-mold were hot rolled into slabs of 1-3 mm thickness, depending on the type of test carried out.

2.2 Microstructural investigation

Samples for microstructural characterization were prepared from the L-shaped mold casting in the as cast condition using 5A composite. The volume fraction and average size of the B_4C particles was measured using Clemex image analyzer. Fracture surfaces were examined of samples sectioned from both tensile- and impact tested bars. The samples were examined using Hitachi S-7000 and Hitachi SU-8000 FE-SEM microscopes equipped with EDS facilities at McGill University, Montreal.

2.3 Mechanical testing

Charpy impact testing was carried out on un-notched test specimens (10x 10 x 55 mm). The samples were sectioned from the L-shaped mold castings and heat treated in an electrical air forced furnace. An instrumented Charpy impact testing machine, equipped with a data acquisition unit was employed to measure the load, total absorbed energy (E_t) to fracture. The mean values of 6 impact-tested samples for each composite/condition were reported.

Slabs (25x 20x 400 mm) were prepared from the book mold castings. Prior to rolling using a four cylinder mill, the slabs were annealed at 500°C for 16 h. The last two passes were carried out at room temperature to straighten the rolled slabs (sheets) – see **Figure 4**. **Figure 5** shows the dimensions of samples prepared from

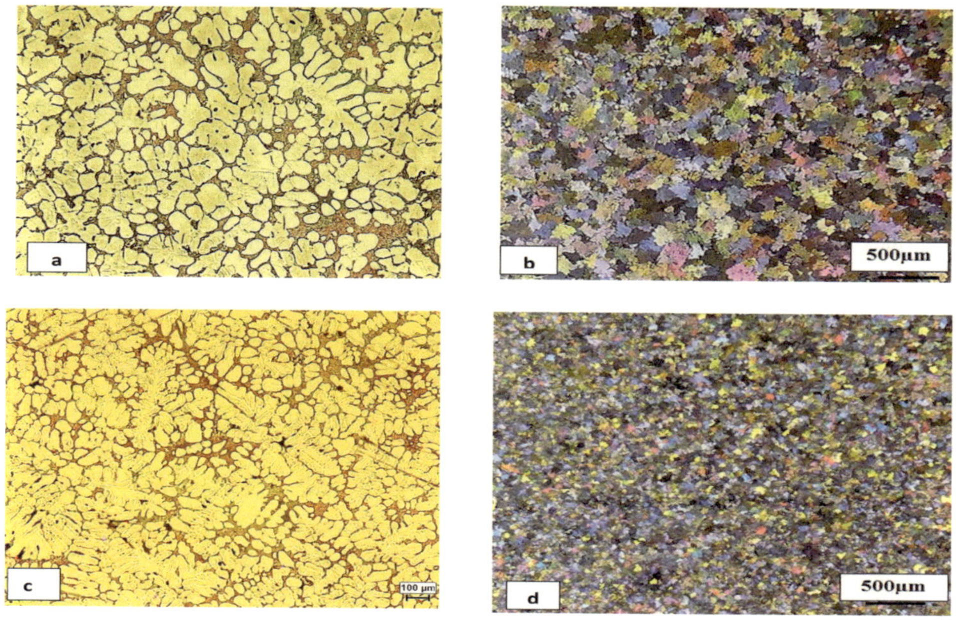

Figure 3.
(a, c) Optical micrographs of Al-7%Si alloy for (a) block casting, 60 µm; (c) L-shaped casting, 30 µm; (b, d) Macrographs showing grain size in (b) block casting; (d) L-shaped casting.

Figure 4.
Hot rolled sheets.

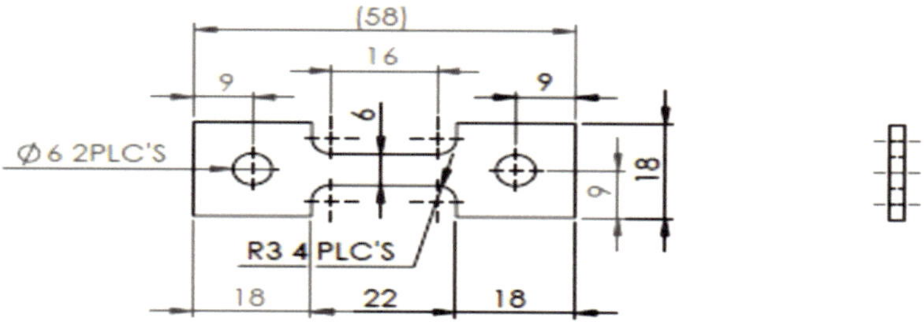

Figure 5.
Typical sample for room and high temperature tensile testing (dimensions are in mm).

the rolled sheets and used for room and high temperature tensile testing. Tensile samples (matrix is aluminum) were solutionised at 620°C for 24 h. In spite of the fact that pure aluminum normally is not heat treatable, it could benefit from the precipitation of Zr-rich particles during aging. The $6063/B_4C/15p$ composite samples were solutionized at 540°C to minimize surface oxidation (MgO). After solution heat treatment, the tensile bars were quenched in warm water (60° C), followed by aging for 10 h at 200, 300 and 400° C, and then air cooling. Room temperature testing was carried out using an MTS Servohydraulic mechanical testing machine at a strain rate of 4×10^{-4}/s.

High temperature testing was done at strain rate of 5×10^{-4}/s in a temperature range 25–500°C. In all cases, tensile properties were measured: ultimate tensile strength (UTS), the 0.2% offset yield strength (YS) and percentage elongation (%El). For each working condition, at least five specimens were tested and mean values were reported (SD ±5%). Microstructure and fracture behavior of selected samples were examined using optical microscopy and Field Emission Scanning Electron Microscopy (FESEM) techniques.

3. Results and discussion

3.1 Microstructural characterization (as cast condition)

The main function of the addition of Zr and Ti, is to protect the B_4C particles from reacting with the molten Al [45–48]. **Figure 6(a)** depicts the microstructure of a specimen sectioned from the L-shaped castings, revealing a uniform distribution of B_4C particles throughout the matrix. From such micrographs, the volume fraction of B_4C particles was determined (~15 vol.%). According to the Al–Ti binary diagram [49], at 730°C an amount of 0.5 wt-%Ti could be added to the molten composite. From the reported findings of Tahiri et al. [50–54] it was reported that increasing the concentration of Ti in the molten alloy above 0.5 wt.% will increase the temperature of the molten alloy. As a result, fluidity of the composite will be markedly reduced caused by the segregation of B_4C particles as exhibited in **Figure 6(b)**.

FESEM examination of 5A composite treated with Ti revealed that in addition to B_4C particles, possible precipitation of several intermetallics mainly, TiB_2, TiC and traces of $AlB_{24}C_4$, Al_4C_3, Al_3BC and AlB_{12}, along with the primary intermetallic phases TiAl, Ti_3Al and $TiAl_3$ could also occur. It is expected that the formation of these phases in layers would lead to an improvement in the adhesion between the matrix and the B_4C reinforcement [21–23]. **Figure 7** displays electron micrographs of alloy B,

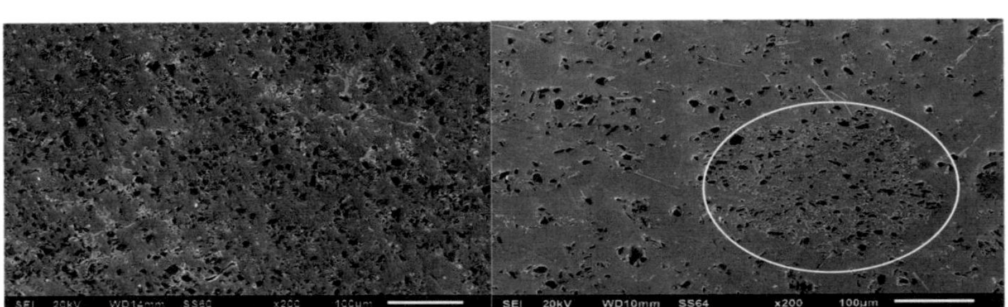

(a) (b)

Figure 6.
Secondary electron micrographs showing: (a) a uniform distribution of B_4C in matrix of base alloy using in-house powder injecting technique, (b) segregation of B_4C (white circle).

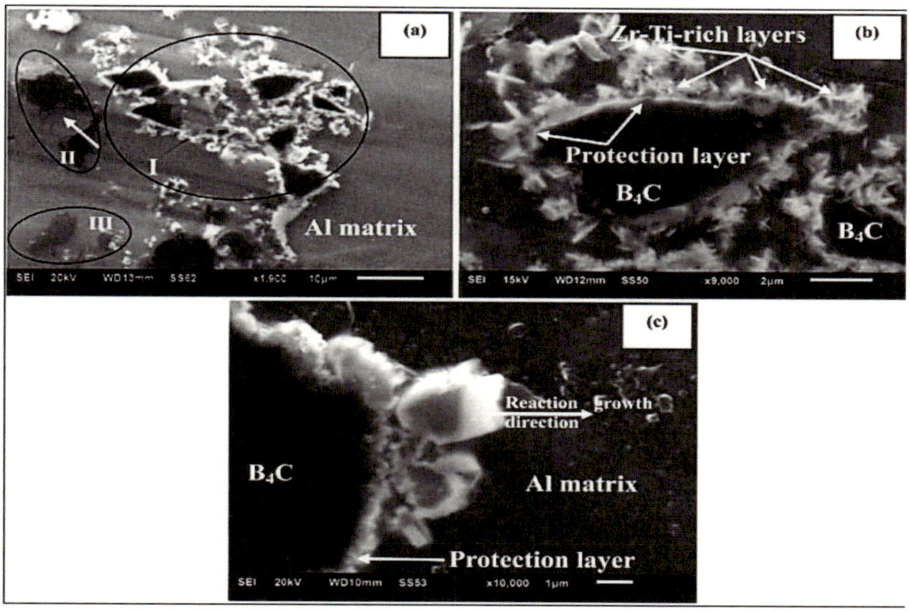

Figure 7.
SE images from composite B showing (a) regular B_4C particles protected by layers of Zr–Ti rica particles, (II) irregular forms of B_4C showing partial reaction with matrix forming AlBC compound; (b) B_4C particle surrounded by two layers of Zr–Ti rich particles; and (c) B_4C particle showing progress of cumulative reaction towards matrix.

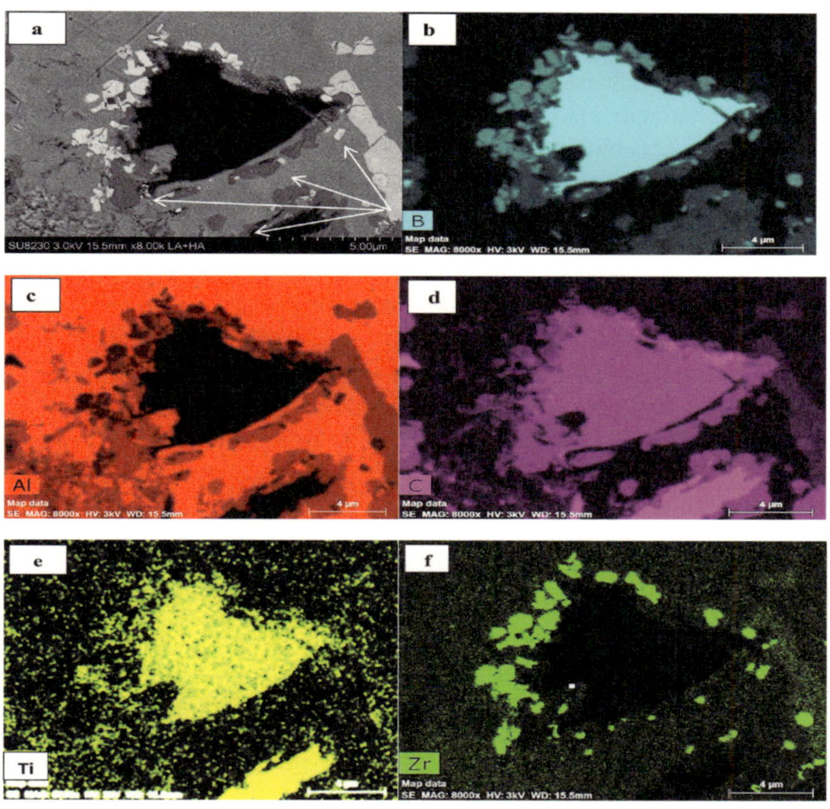

Figure 8.
Element distribution: (a) backscattered electron image-note formation of several particles around a B_4C particle, white arrows, (b) boron, (c) alumium, (d) carbon, (e) titanium, (f) zirconium.

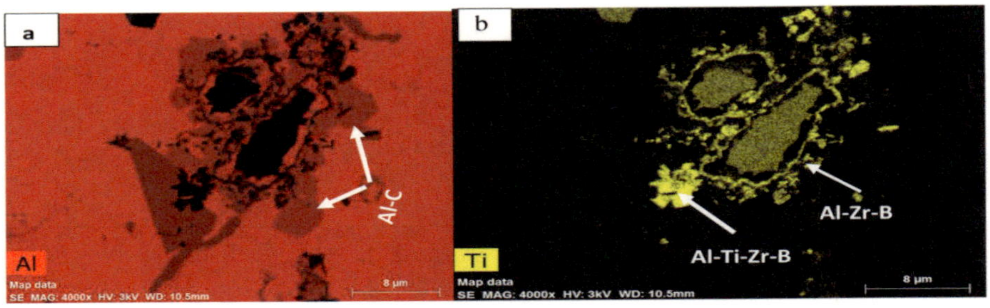

Figure 9.
*B_4C-matrix interactions: (a) formation of Al-C, (b) formation of Zr rich phases-see **Figure 7(c)**.*

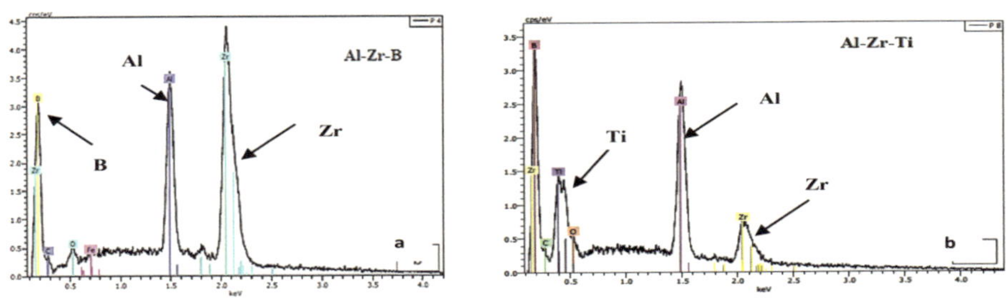

Figure 10.
*EDS spectra obtained from **Figure 9(b)** confirming the B4C-matrix interaction and the dependence of the composition of outcome on its position with respect to B4C particles.*

containing Ti and Zr. As can be seen in **Figure 7(a)**, the B_4C particles are surrounded by several layers of Zr–Ti rich phases (area marked I). Area marked II shows B_4C particles that have partially reacted with the matrix due to formation of the layer of AlBC existing in the matrix. In area III, fine B_4C particles are found be transformed completely into AlBC compounds. **Figure 7(b)** reveals a B_4C particle surrounded by a thin layer of Ti-rich phase followed by several layers of Zr–Ti rich phases. Some of these Zr–Ti rich phase particles are seen to grow into the aluminum matrix.

Figure 8 is produced from 5A composite remelted for multiple times at 730°C. **Figure 8(a)** shows the distribution of the B_4C particles. The B, Al and C distribution are presented in **Figures 8(b), 8(c), 8(d)**, respectively. Another point to be considered is that Ti covers the entire surface of the B_4C particle (similar to C and B) whereas Zr is limited to the layer decorating the B_4C particle. It is inferred from **Figure 9** that the layers surrounding the B_4C particles are a mixture of Al-Ti, and Al-C-Ti compounds. The EDS spectra obtained from **Figure 9(b)** are presented in **Figure 10**. Based on these EDSs, areas near the B_4C particles could be made of Al-B-Zr compound whereas those away are probably Al-Ti-Zr compound [55].

4. Mechanical properties

4.1 Impact testing

Total energies (Et) produced from the ten studied composites in the solutionized condition are shown in **Figure 11(a)**. Apparently the absorbed energy of the composite depends on what matrix is used and the volume fraction of the undissolved intermetallics. Following aging at 200°C for 10 h (**Figure 11(b)**), the precipitation of Zr-and Sc containing phases [55–58] led to significant decrease in the values of Et which may be attributed to precipitation of Mg_2Si phase particles

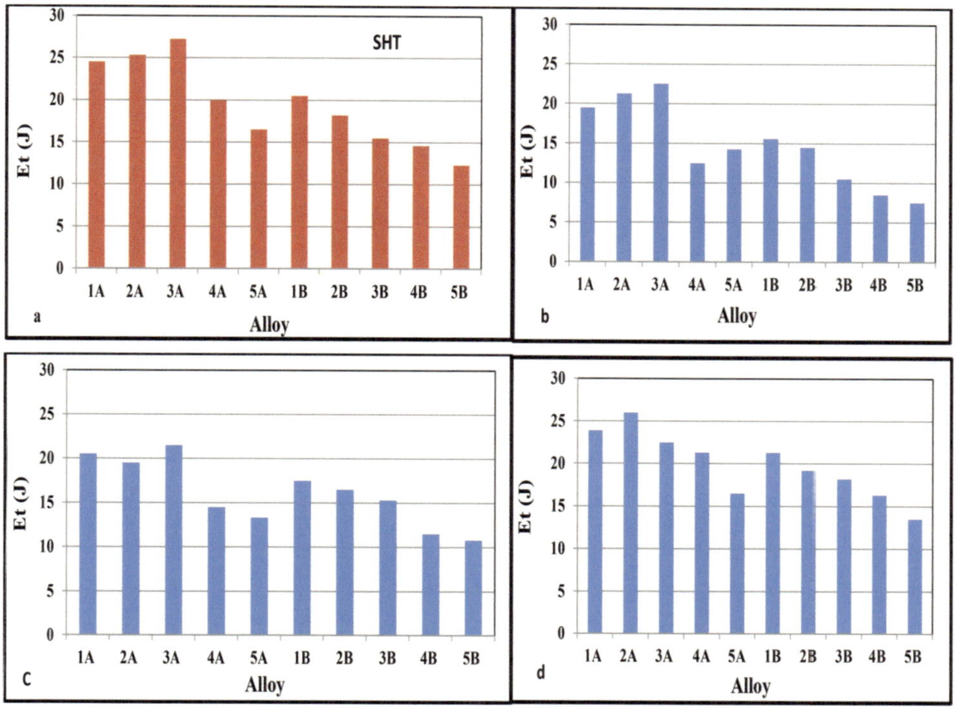

Figure 11.
Total absorbed energies of the present composites: (a) SHT, (b) aging at 200°C, (c) aging at 300°C, (d) aging at 400°C.

during aging in particular in B-series. It is inferred from **Figure 11(b)** that precipitation of Zr and/or Sc phases has an insignificant effect on the absorbed energy in series A-composites. According to Fuller et al. [59] aging 6063 alloy at 300°C would result in alloy softening due to coarsening of Mg_2Si phases particles. Simultaneous precipitation of Zr-rich phase may lead to balancing the composite toughness to some extent (**Figure 11(c)**). Aging at higher temperatures i.e. 400°C for 10 h resulted in coarsening of all types of precipitated phases causing important improvement in the composite toughness, regardless of the type of the matrix used, as exhibited in **Figure 11(d)**.

Fracture mechanism of Al-2%Cu composite was investigated by Miserez [60]. The study showed that the fracture may occur in two stages: (i) particle fracture leading to void nucleation in the matrix, and (ii) voids nucleated in the matrix in areas of high stress concentrations. The blue arrow in **Figure 12(a)** shows that the crack is propagating through the protecting layer surrounded by stacking faults (white arrows). On the left hand side of the micrograph several stacking faults appear in the form of steps (white arrows). Two distinctive types of cracks were observed in **Figure 12(b)**: cracks that took place at the interior of the B_4C particles and continued through the protecting layers i.e. intergranular, or those occurring at the B_4C/matrix interfaces (black arrow). No particle debonding was observed due to the existence of the protecting layers as displayed in **Figure 12(c)**. The microstructure beneath the fracture surface (vertical section-loading direction) shown in **Figure 12(d)** demonstrates the coherency between the B_4C particles and the surrounding matrix. Aging the composite at 400°C led to marked coarsening of the Al_3Zr phase particles as seen in **Figure 13(a)**, which explains the improvement in the composite toughness in **Figure 11(d)**. The EDS spectrum in **Figure 13(b)** corresponds to the circled area in **Figure 13(a)** revealing strong reflections from Al and Zr elements.

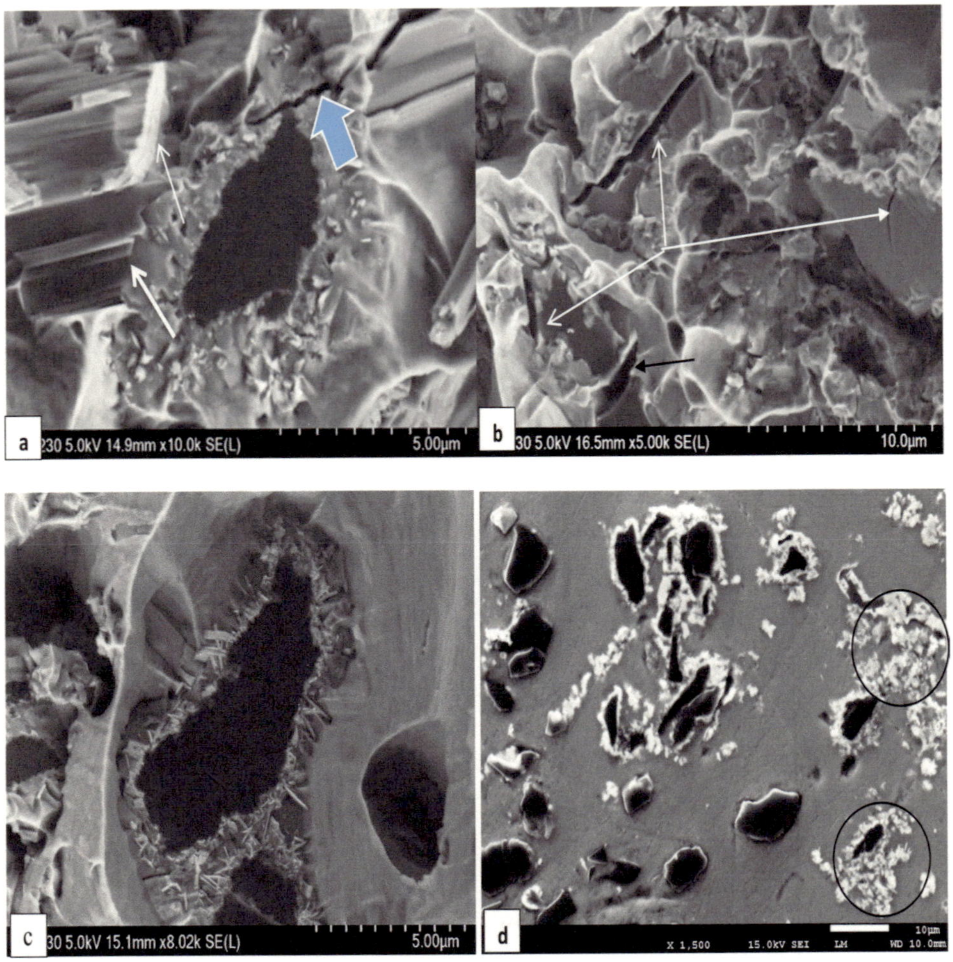

Figure 12.
Fracture characteristics of the present composites: (a) stacking faults-SHT, (b) cracks- aging at 200°C/10 h, (c) B_4C/matrix coherency, (d) vertical section beneath (c) confirming particle/matrix bonding-note the severe reaction around some of the B_4C particles - circled areas.

4.2 Tensile testing

4.2.1 Room temperature testing

The stress–strain curves of two aluminum matrix composites in the SHT condition and after aging at 200°C/h are shown in **Figure 14**. The main observation to be made is the slow working hardening rate illustrated by low work hardening and the slow increase in the composite UTS (**Figure 14(a)**). As a consequence of aging at 200°C/10 h, the UTS increased by approximately 80 MPa which may be attributed to the precipitation of Al_3Zr phase particles [61]. Considering the solutionizing treatment of 5B composite the maximum attainable strength is about 280 MPa- **Figure 14(b)**. Using a heat treatable matrix i.e. 6063 alloy, the composite revealed significant improvement in both the UTS levels as well as work hardening rate as displayed in **Figure 14(c)** which may be caused by the precipitation of Mg_2Si phase particles during the storing period prior to testing (~10 minutes at room temperature). As expected, aging at 200°C/10 h resulted in increasing the composite strength from 280 MPa to 500 MPa, **Figure 14(d)**, which may be interpreted in terms of simultaneous precipitation of both Mg_2Si and Al_3Zr phase particles.

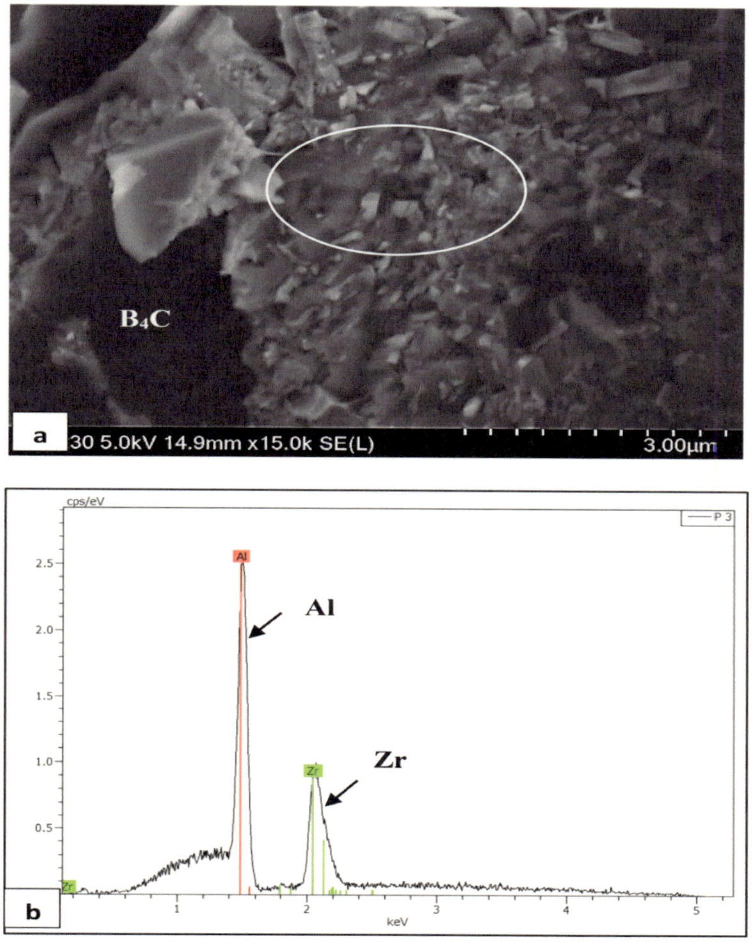

Figure 13.
(a) Fracture surface of composite 5B aged at 400°C/10 h, (b) EDS spectrum corresponding to white circle in (a).

Following the solution heat treatment of composite 5A, the fracture surface is characterized by the formation of deep dimple network as demonstrated in **Figure 15(a)**. Some of these dimples revealed the presence of deformation bands (arrowed) due to composite ductility. The marking seen on the surface of the B_4C particles in **Figure 15(b)** may be caused by gradual fracture of the particles, maintaining at the same time their coherency with the aluminum matrix. Precipitation of Al_3Zr phase particles during aging at 200°C/10 h is clearly seen in **Figure 15(c)**. Due to reduction in the composite ductility, some of the B_4C particles were cracked as shown by the white arrows in the same figure. In the case of 6063 alloy matrix, with the significant increase in the composite UTS level following aging at 200 C/10 h (500 MPa), cracks are seen to initiate and propagate through the Zr-Ti protecting layer as demonstrated in **Figure 15(d)** – see blue arrow. No B_4C particle debonding is observed to take place under axial loading.

4.2.2 High temperature testing

The 5A and 5B composites (**Table 1**) were tested in the temperature range of 25–500°C and their corresponding stress–strain curves are displayed in **Figure 16**. Composite 5B showed a slightly higher strength compared to composite 5A. It should

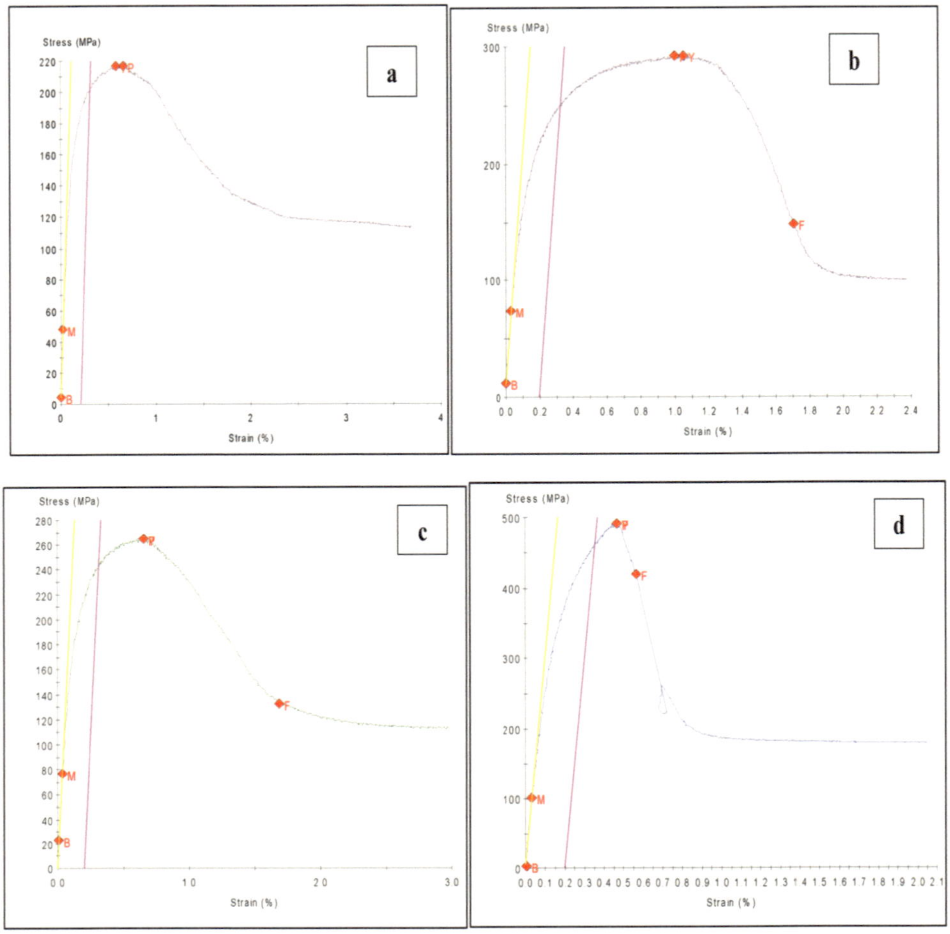

Figure 14.
Stress–strain diagrams corresponding to: (a) Al/B_4C 5A composite - SHT, (b) Al/B_4C composite aged at 200°C/ 10 h, (c) 5B 6063/B_4C composite - SHT, (d) 6063/B_4C composite aged at 200°C/ 10 h.

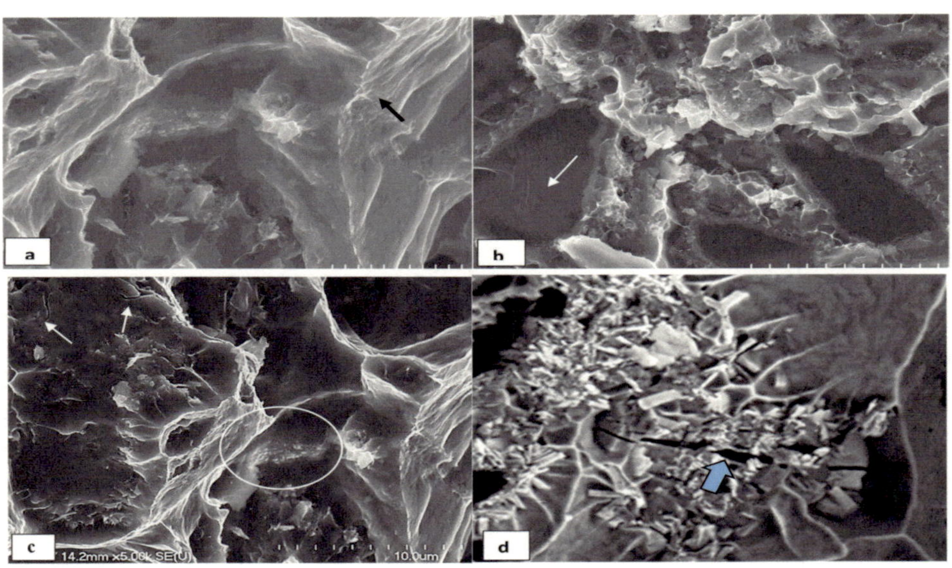

Figure 15.
Fracture surface of tensile tested samples of 5A composite: (a,b) SHT, (c,d) aging at 200 C/10h.

be mentioned here that samples of composite 5B were tested in the T4 condition which involves natural aging. Increasing the testing temperature up to 450°C resulted in significant increase in the composite pct. Elongation to failure. Aging at further higher temperature would lead to precipitation of Al_3Zr which would result in reducing the composite ductility, as shown in **Figure 16(c)** [6, 62–65].

For aging at temperatures higher than 0.5 of the melting temperature (Tm), there is a similarity between creep and hot deformation. Under this condition, the relation between the measured parameters can be expressed using power law relationships as described by Eq. 1 [66–70]:

$$\dot{\varepsilon}\exp\left(\frac{Q_a}{RT}\right) = A\sigma^n = Z \quad (1)$$

where $\varepsilon^{\cdot}$ = strain rate, σ = flow stress, n = stress exponent, Q_a = activation energy, R = gas constant, T = absolute temperature, A is a constant and Z = Zener-Hollomon parameter. At a constant $\varepsilon^{\cdot}$, σ^n can expressed as:

$$\sigma^n = B\exp\left(\frac{Q_a}{RT}\right) \quad (2)$$

where B = constant. Differentiation of Eq. 2 coupled with (1/T), gives Q_a as:

$$\frac{\partial \ln\sigma}{\partial\left(\frac{1}{T}\right)} = \frac{Q_a}{nR} \quad (3)$$

Applying these equations, the plot of lnσ *vs* $1/T$, will give a straight line with a slope of (Q_a/nR) as shown in **Figure 17** [66–68].

The fracture surface of composite 5A tested at 25°C was characterized by the presence of deformation bands covering the internal surface of the dimples as shown previously. **Figure 18(a)** exhibits the fracture surface of 5A composite tested at 250°C revealing multiple contour-type markings (white arrow) due to the high ductility ~15%. Testing at 350°C resulted in major increase in the deformation bands in the form of steps (blue arrows) as shown in **Figure 18(b)**. The thin white arrows point to cracked B4C particles. The black arrow in **Figure 18(c)** - 5A composite pulled to fracture at 450°C- indicates the presence of a long crack within the protective layer. In addition, the 5A composite exhibited elongated dimples as depicted in **Figure 18(c)**. **Figure 18(d)** is an enlarged portion of the crack in **Figure 18(c)**.From the associated EDS spectrum in **Figure 18(e)**, the possibility of precipitation of a large amount of Zr-rich particles, which would explain the reduction in the composite ductility when tested at this temperature. Fractographic observations made by Zhang et al. [69] on 6092/(B_4C)p indicated the possibility of several interfacial bonding characteristics such as good bonding (extruded composites) and weak bonding (hot isostatic pressing). The fracture surfaces of composites would also show a mixture of ductile and brittle types of fracture [70].

The fracture surface of 5B composite pulled to fracture at 250°C revealed that in addition to the deep dimples, some stacking faults could also be seen in the fracture surface (**Figure 19(a)**-white arrows). As in the case of 5A composite, at 350°C, the fracture surface exhibited a well-defined dimple structure as a result of the increase in the composite % elongation to fracture, **Figure 19(b)**-black arrows point to the thickness of the protection layer. Due to the strong particle/matrix interface adhesion, some of the B_4C particles have been cracked at their interior. In this case, the crack was initiated at the particle/matrix interface and propagated through the particle. When the sample was tested at 450°C, the fracture surface revealed the formation of very large and deep dimple network as shown in **Figure 19(c)**.

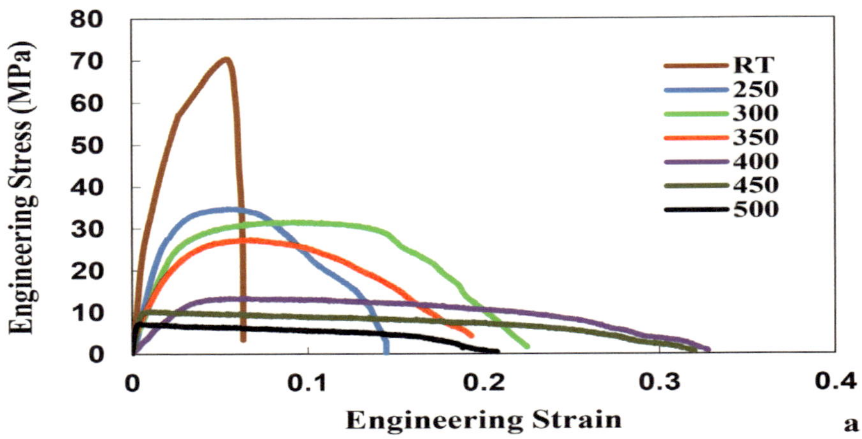

(a)

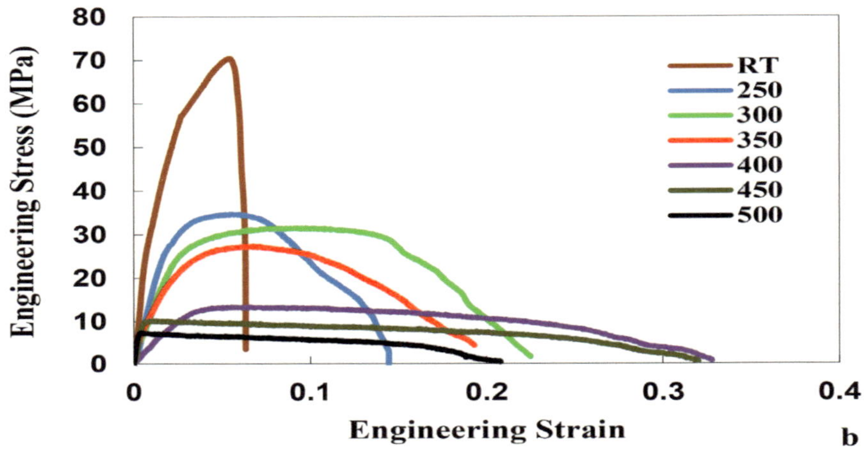

(b)

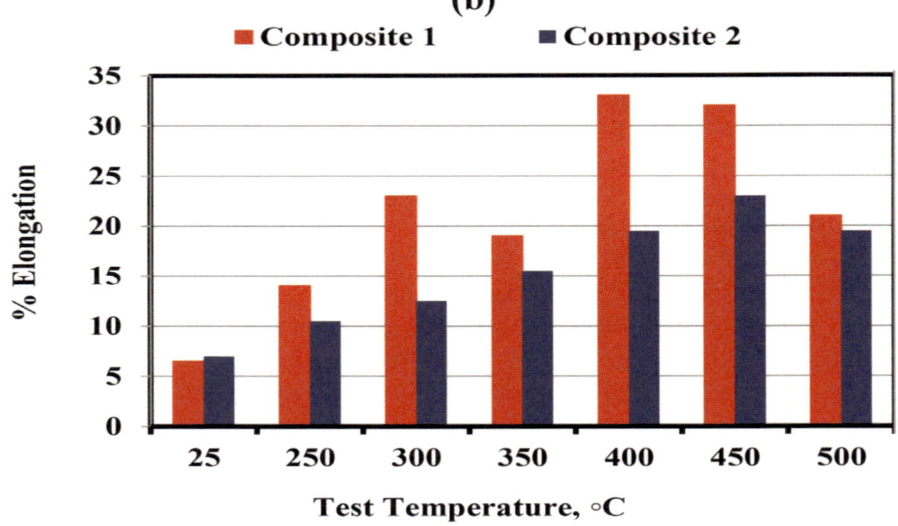

(c)

Figure 16.
(a) Stress–strain curves obtained from 5A composite, (b) stress–strain curves obtained from 5B composite, (c) % elongation as a function of testing temperatire (composite 1 is 5A, composite 2 is 5B).

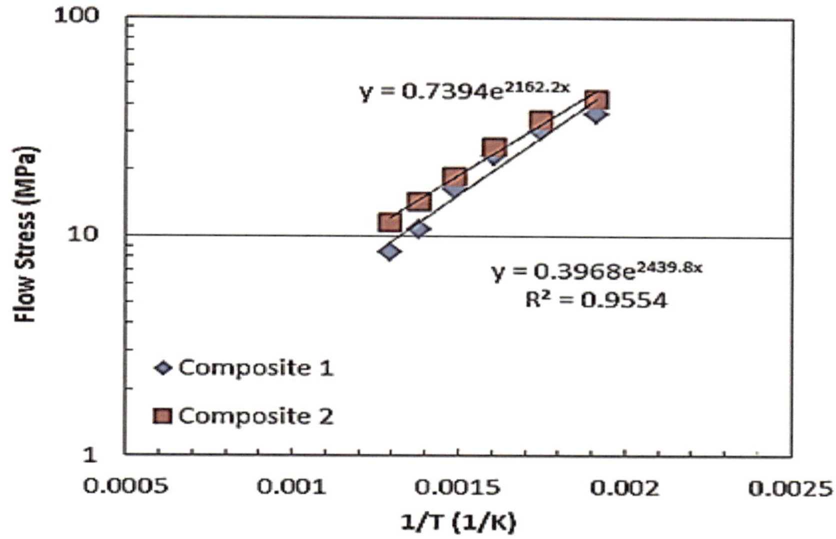

Figure 17.
Relationship of flow stress vs 1/T (T = in kelvin degree).

4.2.3 Effect of strain rate

The main characterized parameters of hot deformation or creep behavior of commercial Al alloys and Al-based composites are by high values of n_a (> 5) as well as the activation energy Q_a. These values are higher than those for solute diffusion [71–75]. This behavior can be explained in terms of interaction of dislocations with the dispersed strengthening particles resulting in a threshold stress σ_0. In this case, the deformation process is related to an effective stress, $\sigma_e = (\sigma - \sigma_o)$ not the applied stress σ. Therefore, equations 1-2 can be rewritten to take into consideration σ_o as follows

$$\dot{\varepsilon}\exp\left(Q_t / RT\right) = Z = A'\frac{Gb}{kT}\left(\frac{\sigma - \sigma_0}{G}\right)^{n_t} \quad (4)$$

where: A = constant, k = Boltzmann's constant, b = magnitude of Burgers vector, G = shear modulus and Q_t = true activation energy.

The true stress–strain curves obtained from testing the 5A composite tested at 300°C (a), 400°C (b) and 500°C (c), are respectively depicted in **Figure 20**. These curves can be divided into three stages; strain hardening where the stress increases with strain until reaches a steady state. Stage 2 represents maximum stress, followed stage 3 where necking takes place leading to failure. Generally speaking, with increasing the strain rate would result in an increase in the flow stress. The effect of testing temperature on the behavior of the stress- strain curves at a constant strain rate of 10^{-3} s^{-1} is displayed in **Figure 20(d)**. Increasing the testing temperature led to an increase in the composite ductility at 500°C and higher strain rates higher than 10^{-2} s^{-1}. The ductility was decreased in temperature range of 350° C–450°C.

The relationship between the strain rate, $\dot{\varepsilon}$ and stress σ, at a constant temperature, is governed by plotting $\dot{\varepsilon}$ *vs.* σ applying a log log scale (**Figure 21**) for different testing temperatures. The results reported in **Figure 20** may suggest that the data points at each testing temperature fall on a straight line with a constant na that increases from 5.8 at 500°C to ~7 at 350–450°C, thereafter to 10.4 at 300°C. The high values of n_a are close to those obtained for commercial aluminum alloys [76–80] and metal matrix composites. As mentioned before, dislocations -second phase

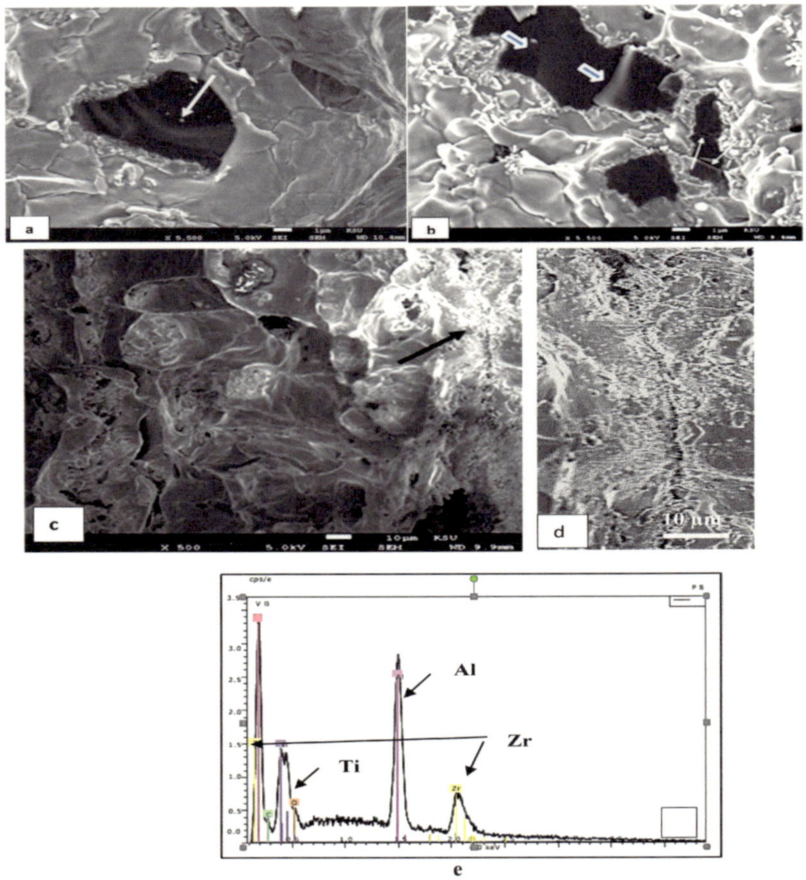

Figure 18.
Fracture surface of tensile tested samples of 5A composite: (a) 250°C, (b) 350°C, (c) 450 C, (d) an enlarged portion of (c) showing the crack, (e) EDS corresponding to (c) revealing reflections due to Al, Zr, Ti elements.

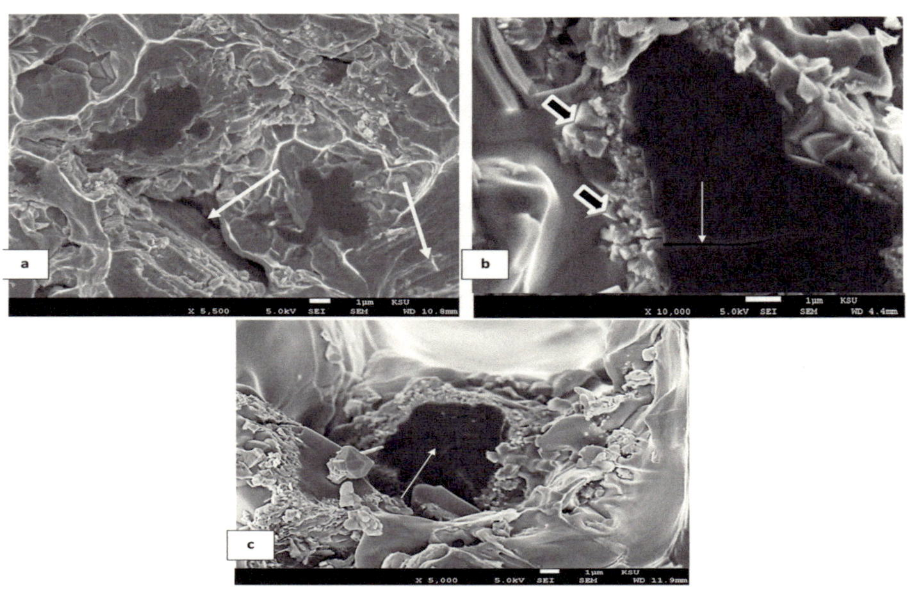

Figure 19.
Fracture behavior of 5B composite tested at: (a) 250°C, (b) 350°C, (c) 450°C.

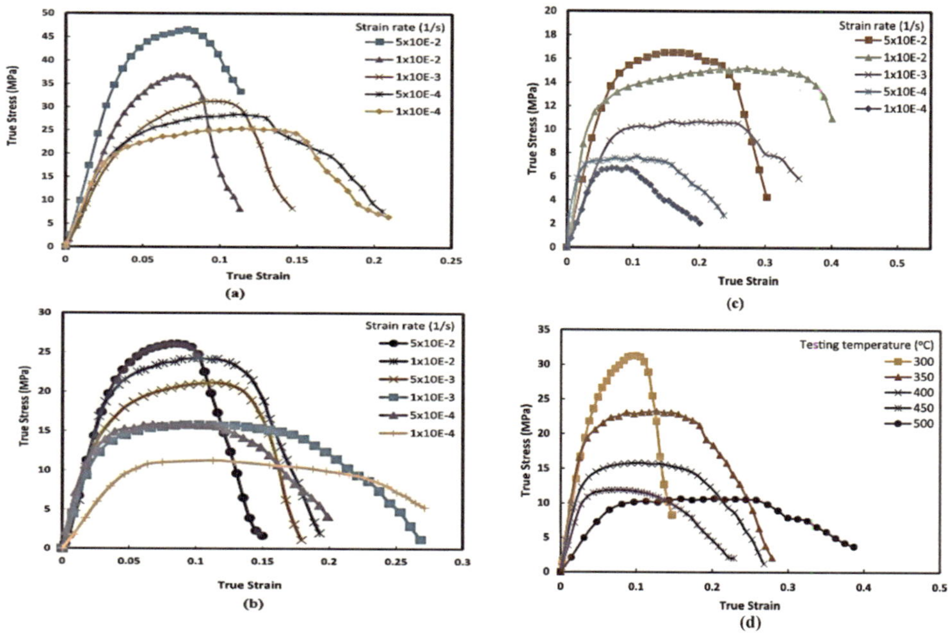

Figure 20.
True stress–strain curves at different strain rates at (a) 300°C, (b) 400°C, (c) 500°C, (d) strain rate $10^{-3}\ s^{-1}$.

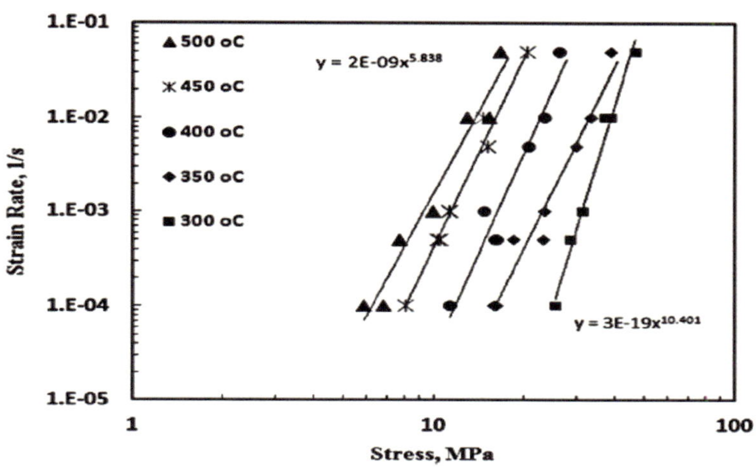

Figure 21.
Strain rate and stress relationship in the temperature range 300–500°C.

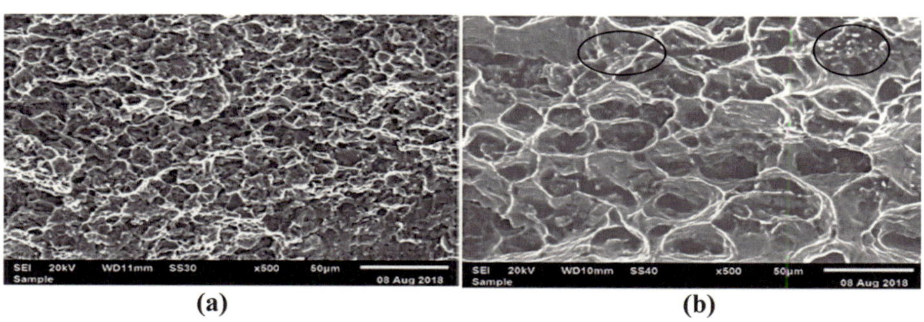

Figure 22.
Fracture surface of samples tested at 300°C: (a) $10^{-2}\ s^{-1}$, (b) $10^{-4}\ s^{-1}$.

particles interaction would lead to high values of n_a and Q_a threshold stress in the composite materials.

Figure 22 (a) shows the fracture surface of the samples tested at 300°C (strain rate of 10^{-2} s^{-1}), consisting of a mixture of small dimples and intragranular fracture. **Figure 22 (b)** is the fracture surface at strain rate of 10^{-4} s^{-1} at 300°C, exhibiting larger dimples with precipitation of Al_3Zr in particles at their interiors (circled areas) [81].

5. Summary

The results obtained from the present investigations revealed that the powder injection technique used in our study proved to be effective in producing composites with a uniform distribution of B_4C particles throughout the matrix (commercial aluminum or 6063 alloy). The combined addition of Zr and Ti improved the possibility of increasing the number of B_4C particles in the matrix by improving the particulate wettability. The precipitating $Al_3(Zr_{1-x}Ti_x)$ particles decorating the B_4C particles were found to grow into the surrounding matrix. Precipitation of Mg_2Si in 6063/B_4C was more effective in controlling the composite toughness than Al_3Zr in the under-aging conditions. Overaging occurred at 400°C for prolonged aging times (i.e. 10 h), resulting in a significant improvement in the composite toughness regardless the type of the matrix. Cracks were always initiated at the particle/matrix interfaces and propagated either through the B_4C particles or along the protecting $Al_3(Zr_{1-x}Ti_x)$ layer. No particle debonding was observed regardless the type of matrix or the testing method. Formation of the Zr/Ti rich layers surrounding the B_4C particles strengthen their adhesion to the surrounding matrix. Increasing the testing temperature leads to rapid decrease in the composite strength in an exponential pattern which appeared in the gradual fracturing of the reinforcement B_4C particles. The plots of flow stress as a function of testing temperature are linear with a fitting factor of 0.955. The value of $n_t \sim 5$ and Q_t of 130 kJ mol^{-1} along with subgrain formation may conclude that dislocation climb is the main controlling process. Similar observation was made in pure Al with dispersed particles. The pct elongation to failure reached a maximum value at intermediate value of Z, which can determine the optimum conditions for the composite formability.

Author details

Mohamed F. Ibrahim[1*], Hany R. Ammar[2], Agnes M. Samuel[1], Mahmoud S. Soliman[3], Victor Songmene[4] and Fawzy H. Samuel[1]

1 Université du Québec à Chicoutimi, Chicoutimi, Québec, G7H 2B1, Canada

2 Department of Mechanical Engineering, Qassim University, Saudi Arabia

3 Department of Mechanical Engineering, King Saud University, Riyadh, Saudi Arabia

4 Department of Mechanical Eng., École de Technologie Supérieure, Québec, Canada

*Address all correspondence to: fhsamuel@uqac.ca

References

[1] K. Kalaiselvan, N. Murugan, S. Parameswaran, Production and characterization of AA6061–B4C stir cast composite, Materials and Design, 32 (2011), pp. 4004-4009.

[2] S. Roy, S. Sharma, S. Sharma, Review on Fabrication of Aluminum 7075+ B4C Composites and its Testing, Int. J. of Advanced Research in Science and Eng., 6(2017), pp. 647-652.

[3] K.E. Öksüz and K.O. Oskay, The Effects of Aging on the Hardness and Wear Behaviour of 2124 Al Alloy/B4C Composites, Proceedings of the 4th International Congress APMAS2014, April 24-27, 2014, Fethiye, Turkey, 127 (2015), pp. 1367-1369.

[4] A. Singla, S. Shandilya, P. Gera, A. Gupta, Process Parameter Optimization using DOE Methodology on Al- MMC to Maximize Mechanical Properties, International Journal of Emerging Science and Engineering (IJESE), ISSN: 2319-6378, Volume-5 Issue-2, January 2018, pp. 4-7.

[5] S. Mohan, E. Kennedy, MACHINABILITY STUDIES OF ALSIB4C DEVELOPED MMC DEVELOPED USING STIR CASTING, Journal of Engineering & Technology, 4(2014), pp. 25-28.

[6] R.U. Vaidya, S.G. Song, A.K. Zurek, Dynamic mechanical response and thermal expansion of ceramic particle reinforced aluminum 6061 matrix composites. Philos Mag A, 70 (1994): pp. 819-836.

[7] Y. Kumar, G. Anil Kumar, J. Satheesh, T. Madhusudhan, A Review On Properties Of Al-B4C Composite Of Different Routes, International Research Journal of Engineering and Technology, 3 (2016), pp. 860-865.

[8] İ. Topcu, Investigation of Wear Behavior of Particle Reinforced AL/B4C Composites under Different Sintering Conditions, TEHNIČKI GLASNIK 14, (2020), pp. 7-14.

[9] B. Manjunatha, H. B. Niranjan, K.G.Satyanarayana, Effect of amount of boron carbide on wear loss of Al-6061 matrix composite by Taguchi technique and Response surface analysis. Materials Science and Engineering, 376(2018), https://doi.org/10.1088/1757-899X/376/1/012071.

[10] K.M. Shorowordi , A.S.M.A. Haseeb J.P. Celis, Tribo-surface characteristics of Al–B4C and Al–SiC composites worn under different contact pressures, Wear, 261 (2006), pp. 634-641.

[11] K.M. Shorowordi, T. Laoui, A.S.M.A. Haseeb, J.P. Celis, L. royen, Microstructure and interface characteristics of B4C, SiC and Al2O3 reinforced Al matrix composites: a comparative study, J.Mater. Process. Technol. 142 (2003), pp. 738-743.

[12] K.M. Shorowordi, A.S.M.A. Haseeb, J.P. Celis, Velocity effects on the wear, friction and tribochemistry of aluminium MMC sliding against phenolic brake pad, Wear 256 (2004)pp. 1176-1181.

[13] G. Straffellini, M. Pellizzari, A. Molinari, Influence of load and temperature on the dry sliding behaviour of Al-based metal-matrix-composites against friction material, Wear 256 (2004), pp.754-763.

[14] M. Sing, D.P. Mondal, O.P. Modi, A.K. Jha, Two-body abrasive wear behaviour of aluminum alloy–sillimanite particle reinforced composite, Wear, 253 (2002),pp. 357-368.

[15] R.M. Mohanty, K. Balasubramaniam, S.K. Seshadri,

"Boron-Carbide Reinforced Aluminium 1100 matrix composites: Fabrications and properties", Materials Science and Engineering A, 498, (2008), pp. 42-52.

[16] A. Canakci "Microstructure and abrasive wear behavior of B4C particle reinforced 2014 Al matrix composites", J. Mater. Sci., 46 (2011),pp. 2805-2813.

[17] I. Kerti and F. Toptan , Microstructural variations in cast B4C-reinforced aluminum matrix composites (AMCs), Mater Lett, 62, (2008). P.1215.

[18] V. Auradia, G.L. Rajesh, S. A. Kori, Processing of B4C Particulate Reinforced 6061Aluminum Matrix Composites by melt stirring involving two-step addition, Procedia Materials Science, 6 (2014), pp.1068-1076.

[19] A.R. Kennedy, B. Brampto, The reactive wetting and incorporation of B4C particles into molten aluminum, Scripta Mater, 44, (2014), pp.1077-1082.

[20] H.O. Topcua, N. Gulsoy A.N. Kadiogluc, Gulluoglua, "Processing and mechanical properties of B4Creinforced Al matrix composites", J. Alloys Compd. 482 (2009), 516.

[21] I. Topcua, H.O. Gulsoy, N. Kadioglu, A.N. Gulluoglu, Processing and mechanical properties of B4C reinforced Al matrix composites, Journal of Alloys and Compounds,482(2009), pp. 516-521.

[22] K. Kalaiselvan, N.Murugan, S. Parameswaran, Production and characterization of AA6061–B4C stir cast composite, Materials & Design, 32(2011), pp. 4004-4009.

[23] M. Uthayakumar, S. Aravindan, K. Rajkumar, Wear performance of Al–SiC–B4C hybrid composites under dry sliding conditions, Materials and Design 47 (2013), pp. 456-464.

[24] Y. Mazaheri, M. Meratian, R. Emadi, A.R. Najarian, "Comparison of microstructural and mechanical properties of Al–TiC, Al–B4C and Al–TiC–B4C composites prepared by casting techniques" Materials Science & Engineering A, 560 (2013),pp. 278-287.

[25] F. Toptan, A. Kilicarslan, I. Kerti, The Effect of Ti Addition on the Properties of Al-B4C Interface: A Microstructural Study, Materials Science Forum, 636-637 (2010), pp. 192-197.

[26] R. Ipek, Adhesive wear behaviour of B4C and SiC reinforced 4147 Al matrix composites (Al/B4C–Al/SiC), J. Mater. Procng. Technol, 162-163 (2005), pp. 71-75.

[27] F. Bedir, Characteristic properties of Al–Cu–SiCp and Al–Cu–B4Cp composites produced by hot pressing method under nitrogen atmosphere, Mater. & Design 28 (2007) 1238-1244.

[28] M. Aizenshtein, N.Froumin, E. Shapiro-Tsoref, M.P. Dariel, N. Frage, Wetting and interface phenomena in the B4C/(Cu–B–Si) system, Scripta Mater. 53 (2005), pp. 1231-1235.

[29] J. Jung and S. Kang, Advances in Manufacturing Boron Carbide–Aluminum Composites, J. Am. Ceram. Soc., 87(2004), pp.47-54.

[30] P. Shen, B. Zou, S. Jin, Q. Jiang, Reaction mechanism in self-propagating high temperature synthesis of TiC-TiB2/Al composites from an Al-Ti-B4C system, Mater. Sci. & Eng. A 454-455 (2007), pp. 300-309.

[31] P.Kumar and R. Parkash, Experimental investigation and optimization of EDM process parameters for machining of aluminum boron carbide (Al–B4C) composite, Machining Science and Technology, 20 (2016), 330-348.

[32] Wang HY, Jiang QC, Wang Y, Ma BX, Zhao F. Fabrication of TiB2 particulate reinforced magnesium matrix composites by powder metallurgy. Mater Lett 58(2004), pp.3509-3513.

[33] H. Yang, T. D. Topping, K. Wehage, L. Jiang, E. J. Lavernia, J. M. Schoenung, Tensile behavior and strengthening mechanisms in a submicron B4C-reinforced Al trimodal composite, Materials Science and Engineering: A, 616 (2014), pp. 35-43.

[34] A. Yazdani and E. Salahinejad, Evolution of reinforcement distribution in Al–B4C composites during accumulative roll bonding, Materials and Design 32 (2011), pp. 3137-3142.

[35] H. Junaedi, M.F.A. Ibrahim, H.R. Ammar, A.M. Samuel, M.S. Soliman, A.A. Almajid, and F.H. Samuel, Effect of testing temperature on the strength and fracture behavior of Al-B4C composites. Journal of Composite Materials, 50(2016): pp. 2871-2880.

[36] M.F. Ibrahim, H.R. Ammar, A.M. Samuel, M.S. Soliman and F.H. Samuel, On the impact toughness of Al-15 vol.% B4C metal matrix composites. Composites Part B: Engineering, 2015. 79 (2015): pp. 83-94.

[37] M.F. Ibrahim, A.M. Samuel, F.H. Samuel, H.R. Ammar and M.S. Soliman, On the Impact Toughness of Al-B4C MMC: The Role of Minor Additives and Heat Treatment, in 119th Metalcasting Congress, AFS 2015, Columbus, OH, April 21-23, 2015. 2015, American Foudry Society: Columbus, Ohio.

[38] M.F. Ibrahim, H.R. Ammar, A.M. Samuel, M.S. Soliman, A.A. Almajid, and F.H. Samuel, Mechanical properties and fracture of Al-15 vol.-%B4C based metal matrix composites. International Journal of Cast Metals Research, 27(2014): pp. 7-14.

[39] M.F. Ibrahim, H.R. Ammar, A.M. Samuel, M.S. Soliman and F.H. Samuel, Neue technologie zur produktion von Al-B4C Metalmatrix-verbundwerkstoffen. Giesserei Praxis, 6(2014): pp. 263-271.

[40] M.F. Ibrahim, A.M. Samuel, H.R. Ammar, M.S. Soliman and F.H. Samuel, A new technology for the production of Al-B4C metal matrix composites. Giesserei-Praxis, 65(2014): pp. 263-271.

[41] M.F. Ibrahim, H.R. Ammar, S.A. Alkahtani and F.H. Samuel, Metallographic investigation of tensile- and impact-tested aluminum composites. Journal of Composite Materials, 2016. 50(20): p. 2793-2805.

[42] M.F. Ibrahim, H.R. Ammar, A.M. Samuel, M.S. Soliman and F.H. Samuel, Metallurgical parameters controlling matrix/B4C particulate interaction in aluminium-boron carbide metal matrix composites. International Journal of Cast Metals Research, 26(2013): pp. 364-373.

[43] M.F. Ibrahim, A.M. Samuel, H.R. Ammar, M.S. Soliman and F.H. Samuel, A New Technology for the Production of Al-B4C Metal Matrix Composites, in Transactions of the American Foundry Society volume 121; 117th Metalcasting Congress ; April 6-9, 2013, [St. Louis, Missouri ; in conjunction with CastExpo'13]. 2013, American Foundry Society: [St. Louis, Missouri; in conjunction with CastExpo'13]. p. 99-110.

[44] M. S. Soliman, M. M. El Rayes, A. T. Abbas, D. Yu. Pimenov, I. N. Erdakov, H. Junaedi, Effect of tensile strain rate on high-temperature deformation and fracture of rolled Al-15 vol% B4C composite, Materials Science & Engineering A, 749(2019), pp.129-136.

[45] A. J. Pyzik, R. A. Newman, A. Wetzel and E. Dubensky: 'Composition control in aluminum boron carbide

composites,' in 'Mechanical properties and performance of engineering ceramics II: ceramic engineering and science proceedings', (ed. R. Tandon et al.), Vol. 27; 2007, Westerville, OH, American Ceramic Society.

[46] X. G. Chen: 'Application of Al–B4C metal matrix composites in the nuclear industry for neutron absorber materials', Proc. On 'Solidification processing of metal matrix composites', San Antonio, TX, USA, March 13-16, 2006 TMS, pp.343-350.

[47] M. C. Flemings: 'Fluidity of metals – techniques for producing ultra-thin section castings', Br. Foundryman, 57(1964), pp.312-325.

[48] M. Rosso, Ceramic and metal matrix composites: Routes and properties, Journal of Materials Processing Technology, 175(2006), pp.364-375.

[49] H. Tahiri: 'Affinement des grains des alliages Al-(0-17%)Si', PhD thesis, Université du Québec à Chicoutimi, Chicoutimi, Que., Canada, 2007.

[50] H. Tahiri, S.S. Mohamed, H.W. Doty, S. Valtierra and F.H. Samuel, Effect of Sr–Grain Refining–Si Interactions on the Microstructural Characteristics of Al–Si Hypoeutectic Alloys. International Journal of Metalcasting, 12(2018): pp. 343-361.

[51] H. Tahiri, A.M. Samuel, H.W. Doty, S. Valtierra and F.H. Samuel, Effect of Sr–Grain Refiner–Si Interactions on the Microstructure Characteristics of Al–Si Hypereutectic Alloys. International Journal of Metalcasting, 12(2018): p. 307-320.

[52] H. Tahiri, S.S. Mohamed, H.W. Doty, S. Valtierra and F.H. Samuel, Effects of Grain Refining on Columnar-to-Equiaxed Transition in Aluminum Alloys, in Aluminum Alloys - Recent Trends in Processing, Characterization, Mechanical Behavior and Applications, S. Sivasankaran, Editor. 2017, InTech.

[53] H. Tahiri, F.H. Samuel and A.M. Samuel, Grain refinement in modified A356 Alloy, in 18th Canadian Materials Science Conference. 2006: McGill University, Montreal, QC, Canada.

[54] H. Tahiri, A.M. Samuel, F.H. Samuel, H.W. Doty and S. Valtierra, Mécanismes d'affinage des grains dans les alliages Al-Si en utilisant des alliages mères, in Int. Symposium on Aluminium : From Raw Materials to Applications, 45th annual Conf. of Metallurgists of CIM, C. 2006, Editor. 2006: Montreal, QC, Canada. pp. 863-880.

[55] K.B. Lee, H.S. Sim, S.Y. Cho Reaction products of Al-Mg/B4C composite fabricated by pressureless infiltration technique. Mater Sci Eng A, A302 (2001): pp. 227-234.

[56] K.E. Knipling, R.A. Karnesky, C.P. Lee, D.C. Dunand, D.N. Seidman, Precipitation evolution in Al-0.1Sc, Al-0.1Zr and Al-0.1Sc-0.1Zr (at. %) alloys during isochronal aging. Acta Mater, 58(2010), pp.5184-5195.

[57] K.E. Knipling, D.C. Dunand, D.N. Seidman, Criteria for developing castable, creep resistant aluminum-based alloys - A review. Z. Metallkunde, 97(2006):pp.246-265.

[58] K.E. Knipling , D.N. Seidman , D.C. Dunand, Ambient- and high-temperature Mechanical properties of isochronally aged Al-0.06Sc, Al-0.06Zr and Al-0.06Sc-0.06Zr (at.%) alloys. Acta Mater, 59 (2011); pp. 943-954.

[59] C.B. Fuller, D.N. Seidman, D.C. Dunand Mechanical properties of Al (Sc, Zr) alloys at ambient and elevated temperatures. Acta Mater, 51(2003):pp. 4803-4814.

[60] A.G.T. Miserez, Fracture and Toughening of High Volume Fraction Ceramic Particle Reinforced Metals. PhD Thesis, École Polytechnique Federale de Lausanne 2003.

[61] H. Zhang, M.W. Chen, K.T. Ramesh, J. Yed , J.M. Schoenung, S.C. Chin, Tensile behavior and dynamic failure of aluminum 6092/B4C composites. Mater. Sci. Eng A, A433(2006):pp.70-82.

[62] X. Kai, Y. Zhao, A. Wang, Hot deformation behavior of in situ nano ZrB2 reinforced 2024Al matrix composites. Compos Sci Technol, 116 (2015):pp. 1-8.

[63] M. Kouzeli and A. Mortensen A. Size dependent strengthening in particle reinforced aluminum. Acta Mater, 50 (2002): 39-51.

[64] M.C. Shankar, J. P. Ka, R. Shettya, Individual and combined effect of reinforcements on stir cast aluminum metal matrix composites-a review. Int J Curr Eng Technol, 3(2013): pp.922-934.

[65] X. Kai, Y. Zhao, A. Wang, Hot deformation behavior of in situ nano ZrB2 reinforced 2024Al matrix composites. Compos Sci Technol, 116(2015):pp. 1-8.

[66] M.E. Kassner, M-T Perez-Prado, Five-power-law creep in single phase metals and alloys. Prog Mater Sci, 45 (2000):pp. 1-102.

[67] M. Abo-Elkhier M.S. Soliman, Superplastic characteristics of fine-grained 7475 aluminum alloy. J Mater Eng Perform 15(2006): 76-80.

[68] M.J. McQueen M.E. Kassner Elevated temperature deformation: Hot working amplifies creep. Mater Sci Eng A, 58 (2005), pp. 410-411.

[69] Zhang H, Ramesh KT and Chin ESC. High strain rate response of aluminum 6092/B4C composites. Mater Sci Eng A, A384 (2004): 26-34.

[70] T.M. Lilo, Enhancing ductility of Al6061-10 wt.% B4C through equal-channel angular extrusion processing. Mater Sci Eng A, A410 (2005): pp.43-446.

[71] F.A. Mohamed, Correlation between creep behavior in Al-based solid solution alloys and powder metallurgy Al alloys, Mater. Sci. Eng. A 245 (1998) 242-256.

[72] E. Evangelista, S. Spigarelli, Constitutive equations for creep and plasticity of aluminum alloys produced by powder metallurgy and aluminum-based metal matrix composites, Metall. Mater. Trans. A 33A (2002) 373-381.

[73] S. Spigarelli, E. Evangelista, S. Cucchieri, Analysis of the creep response of an Al–17Si–4Cu–0.55Mg alloy, Mater. Sci. Eng. A 387-389 (2004) 702-705.

[74] F.A. Mohamed, K.T. Park, E.J. Lavernia, Creep Behavior of Discontinuous SiC-Al Composites, Mater. Sci. Eng. A 150 (1992) 21-35.

[75] R. Fernandez, G. Gonzalez-Doncel, Creep behavior of ingot and powder metallurgy 6061Al, J. Alloy Compd. 440 (2007) 158-167.

[76] K.T. Park, F.A. Mohamed, Creep strengthening in a discontinuous SiC-Al composite, Metall. Mater. Trans. A 26 (1995) 3119-3129.

[77] S. Spigarelli, E. Evangelista, H.J. McQueen, Study of hot workability of a heat treated AA6082 aluminum alloy, Scr. Mater. 49 (2003) 179-183.

[78] E.A. El-Danaf, A.A. Almajid, M.S. Soliman, High-temperature deformation and ductility of a modified 5083 Al alloy, J. Mater. Eng. Perform. 17 (2008) 572-579.

[79] E.A. El-Danaf, A.A. Almajid, M.S. Soliman, Hot deformation of

AA6082-T4 aluminum alloy, J. Mater. Sci. 43 (2008) 6324-6330.

[80] E.A. El-Danaf, M.S. Soliman, A.A. Almajid, Effect of solution heat treatment on the hot workability of Al-Mg-Si Alloy, Mater. Manuf. Process. 24 (2009) 637-643.

[81] S. Spigarelli, C. Paoletti, A new model for the description of creep behavior of aluminum-based composites reinforced with nanosized particles, Compos. Part A:Appl. Sci. Manuf. 112 (2018) 346-355.

Chapter 4

Applications of Rare Earth Metals in Al-Si Cast Alloys

Mohamed Gamal Mahmoud, Yasser Zedan, Agnes-Marie Samuel, Victor Songmene, Herebert W. Doty and Fawzy H. Samuel

Abstract

The present article reviews a large number of research publications on the effect of mischmetal (MM), rare earth metals (RE), La or Ce, and combinations of La + Ce on the performance of Al-Si cast alloys mainly 319, 356, 380, 413, and 390 alloys. Most of these articles focused on the use of rare earth metals as a substitute for strontium (Sr) as a eutectic silicon (Si) modifier if added in low percentage (< 1 wt.%) to avoid precipitation of a significant amount of insoluble intermetallics and hence poor mechanical properties. Other points that were considered were the affinity of RE to react with Sr., reducing its effectiveness as modifier, as well as the grain refining efficiency of the added RE in any form. None of these articles mentioned the exact composition of the RE used and percentage of tramp elements inherited from the parent ore. Using high purity La or Ce proved to have no effect on the Si shape, size or distribution, in particular at low solidification rates (thick sections). However, regardless the source of the RE, its addition to Sr-modified alloys reduced the modification effect. As for grain refining, apparently a high percentage of RE (> 1 wt.%) is required to achieve a noticeable reduction in grain size, however at the cost of alloy brittleness.

Keywords: aluminum alloys, mischmetal, rare earth metals, La, Ce, modification, grain refining, mechanical properties

1. Introduction

Silicon is characterized by its low density ((2.34 g cm^{-3}) coupled with its high hardness (6.5-mohs scale) and low solubility in aluminum matrix which would enhance the alloy wear resistance. The eutectic temperature of Al-11.7% Si alloys is near 577 °C as shown in **Figure 1**, which makes Al-Si alloys easy to cast using different techniques [1–5]. In general, alloys containing Cu and Mg are hardened applying a suitable heat treatable cycle that depends on the thickness of the casting [6–10]. During solidification, the liquid moves along the liquidus line, thus increasing the amount of aluminum, **Figure 2(a)**. At the eutectic temperature, **Figure 2 (b)**, almost 50% of liquid has been solidified. As the temperature continues to decrease (577 °C), the rest of the liquid decomposes into solid Al mixed with solid Si but on a finer scale as presented in **Figure 2(c)**.

Soundness of the cast component depends mainly on the ability of the liquid metal to feed interdendritic regions [11]. In the event of low fluidity, shrinkage cavities may occur, as shown in **Figure 3** [12–14], and in the case of automotive

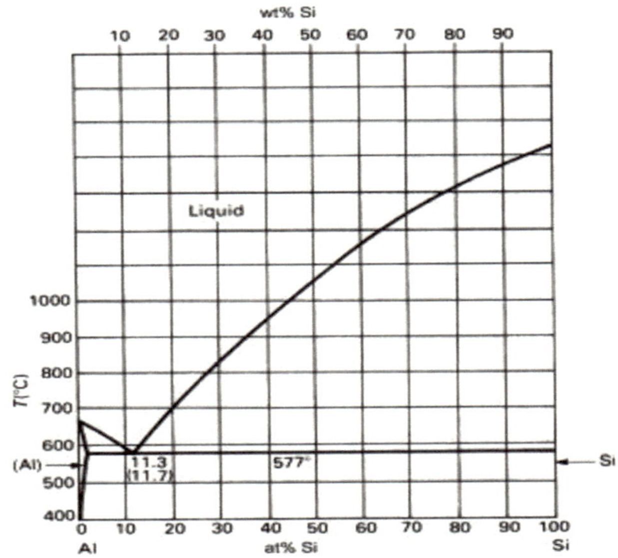

Figure 1.
Al-Si binary diagram [1].

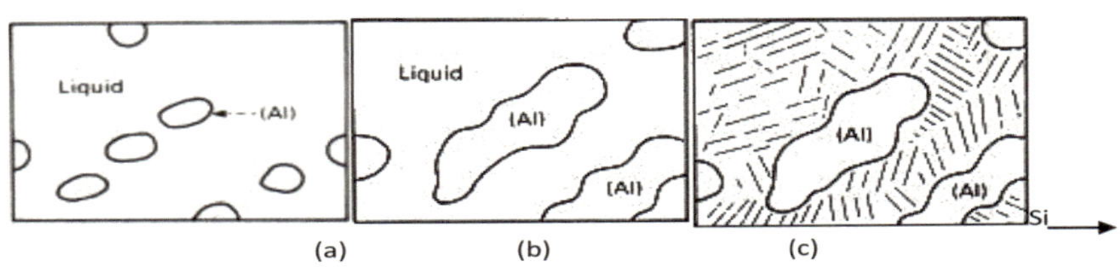

Figure 2.
Progress in liquid to solid during solidification: Start, (b) at eutectic temperature, (c) room temperature [2].

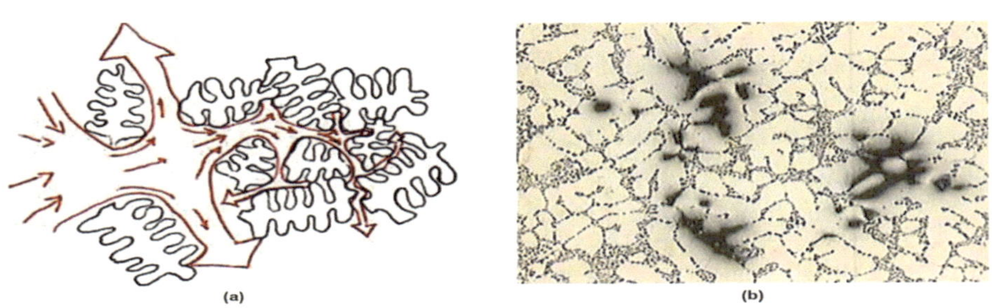

Figure 3.
(a) Schematic diagram showing movement of liquid metal around the dendritic structure, (b) formation if shrinkage cavities [11].

components, rendering the casting poor pressures tightness. Another point to consider is the introduction of tangled oxide films (bifilms) as shown in **Figure 4**.

Table 1 shows the classification of aluminum cast alloys [14] depending on the main alloying element. The number following the decimal point indicates if the alloys are in the form of castings (.0) or ingots (.1 or .2), whereas **Table 2** depicts the different heat treatment designations for these alloys [15, 16]. The composition of Al-Si alloys that are commonly used in aluminum automotive industries is listed in **Table 3** [17].

High entropy alloys (HEAs) are currently receiving much attention in materials engineering because they have potentially desirable properties [18]. Alloys

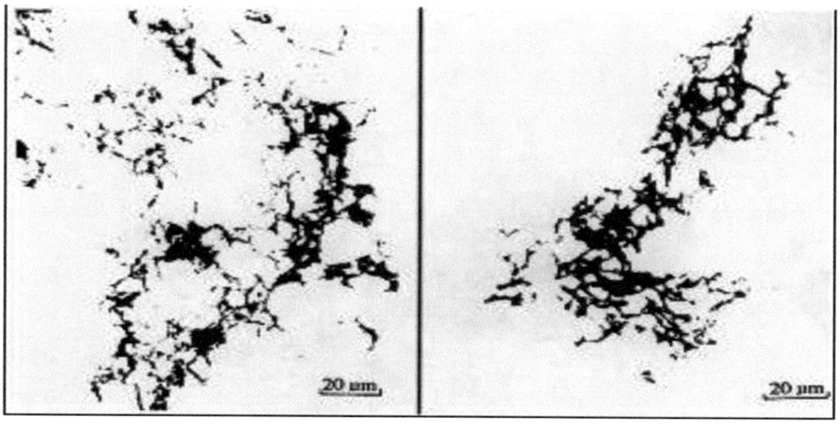

Figure 4.
Examples of tangled oxide films in aluminum alloys [13].

Alloy Series	Principal Alloying Element
1xx.x	99.000% minimum Aluminum
2xx.x	Copper
3xx.x	Silicon Plus Copper and/or Magnesium
4xx.x	Silicon
5xx.x	Magnesium
6xx.x	Unused Series
7xx.x	Zinc
8xx.x	Tin

Table 1.
Classification of aluminum cast alloys.

Letter	Details
F	As fabricated – no heat treatment
O	Annealed – Applies to product which has been heated to produce the lowest strength condition to improve ductility and dimensional stability
H	Strain Hardened – Applies to products which are strengthened through cold-working.
W	Solution Heat-Treated
T	Age hardened alloys - To produce stable other than F, O, or H. The "T" is always followed by one or more digits

Table 2.
Basic heat treatment designations.

containing at least 4–5 elements (5–35 at.% concentrations) are considered as HEAs [19]. Research indicates that some HEAs have considerably better strength-to-weight ratios, with a higher degree of fracture resistance, tensile strength, as well as corrosion and oxidation resistance than conventional alloys. The research investigations on the Al-Si alloys being reviewed here also focus on achieving the same characteristics.

In this review, the microstructures and mechanical properties of aluminum alloys containing different rare earth element additions are discussed, mainly 319,

Alloy	Elements (wt. %)						
	Method(b)	Si	Cu	Mg	Fe	Zn	Others
319.0	S, P	6.0	3.5	<0.10	<1.0	<1.0	
332.0	P	9.5	3.0	1.0	1.2	1.0	
355.0	S, P	5.0	1.25	0.5	<0.06	<0.35	
A356.0	S, P	7.0	<0.20	0.35	<0.2	<0.1	
A357.0	S, P	7.0	<0.20	0.55	<0.2	<0.1	0.05 Be
380.0	D	8.5	3.5	<0.1	<1.3	<3.0	
383.0	D	10.0	2.5	0.10	1.3	3.0	0.15 Sn
384.0	D	11.0	2.0	<0.3	<1.3	<3.0	0.35 Sn
390.0	D	17.0	4.5	0.55	<1.3	<0.1	<0.1 Mg
413.0	D	12.0	<0.1	<0.10	<2.0	—	
443.0	S, P	- 5.25	<0.3	<0.05	<0.8	<0.5	

S = sand casting, P = permanent mold casting, D = die casting.

Table 3.
Composition of Al-Si based cast alloys commonly used in automotive components [17].

356, 380, 411, and 390 alloys which constitute alloys widely used in automotive components. A number of scientific investigations are reported in the literature on the effects of rare earth elements and mischmetal (MM), which is a mixture of rare earth (RE) elements found in abundance in nature with Cerium (Ce) and Lanthanum (La) together comprising approximately 90% of mischmetal. The RE metals investigated include Cerium (Ce), Lanthanum (La), Yttrium (Y) [20–25], Erbium (Er) [26–30], Neodymium (Nd) [31–35], Ytterbium (Yb) [36–40], Samarium (Sm) [41–42], Scandium (Sc) [43–47], Europium (Eu) [48, 49], and Gadolinium (Gd) [50]. Among these, the effects of MM, Ce, La and mixed additions of Ce and La are reviewed.

1.1 Effects of Mischmetal (MM) additions

The effects of small amounts of mischmetal (MM) on the dendrite arm spacing, and the Brinell hardness of Al-1.0% Mg-0.5% Si alloy were investigated by Young et al. [51]. Their results showed that in the range of 0.5–4.0 wt.% MM the hardness increased by more than 30% and the dendrite arm spacing decreased from 50 μm to 18 μm. Ravi et al. [52] analyzed the mechanical properties and microstructure of cast Al-7Si-0.3 Mg (LM25/356) alloy and reported that addition of above 1.0 wt.% of MM lowered the YS, UTS and percent elongation, with an increase in the Brinell hardness. The mechanical properties decreased due to the formation of Ce and La hard intermetallic compounds in the matrix and consumption of a certain amount of Mg in their formation, which reduced the strengthening constituent Mg_2Si formed, contributing to the observed decrease. The yield strength (YS), ultimate tensile strength (UTS), and pct elongation of the Al-7Si-0.3 Mg alloy (in the T6 condition) decreased with the increase in Fe content (from 0.2 to 0.6 pct), as shown in **Figure 5**.

The microstructure and thermal analysis of Al-21 wt.% Si alloys with MM addition were discussed by Chang et al. [53, 54]. According to the authors, addition of 2.0 wt.% MM leads to a morphological change in the primary Si crystals from star-like to polyhedral shape [55] as displayed in **Figure 6**. The thermal analysis results

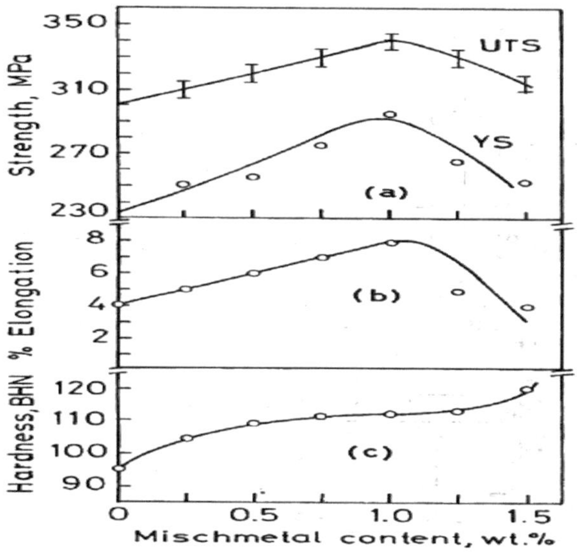

Figure 5.
Mechanical properties of the LM 25 alloys (−T6 condition) containing Fe and Mischmetal [52].

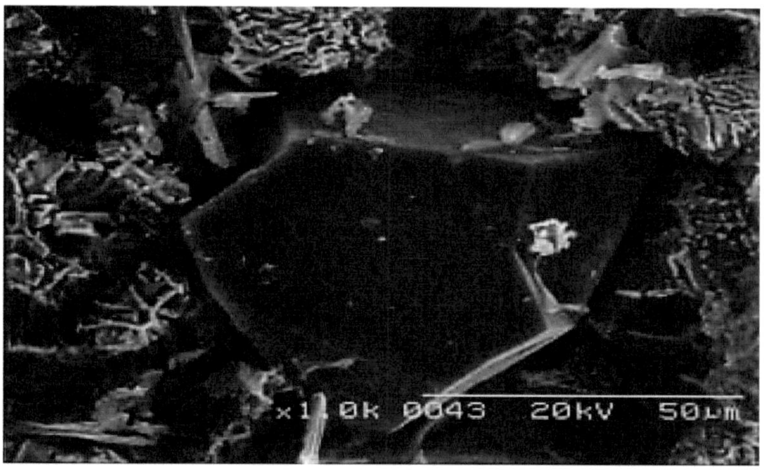

Figure 6.
Scanning electron micrograph of the deep etched Al-21% Si-3% RE alloy. The RE modified primary Si shows typical polyhedral shape [54].

revealed that addition of 3.0 wt.% of MM leads to depressions of 12-17 °C in the primary Si reaction temperature and 2–7 °C in the eutectic Si temperature. Increasing the level of MM additions to in situ Al-15Mg_2Si composite alloy leads to: (i) a reduction in the size of Mg_2Si particles, (ii) a change in the morphology of eutectic Mg_2Si from fibrous to flake like, and (iii) formation of RE-containing compounds in the form of $Al_{11}RE_3$ [56].

Ravi et al. [57] studied the effect of 1.0 wt.% MM additions on the microstructural characteristics and the room and elevated temperature tensile properties of Al-7Si-0.3 Mg (LM 25/356) alloy containing excess iron up to 0.6 wt.%. The results showed that:

i. Alloys with Fe contents ranging from 0.2 to 0.6 wt.%, exhibit grain refinement and partial modification of the eutectic silicon and the finer intermetallic compounds formed with Ce, La, and other elements, thereby improving the strength as well as the ductility of the alloy relative to the same alloys without MM addition.

ii. Alloys containing 0.6 wt.% Fe led to the formation of fine and fibrous shaped intermetallic compounds containing Fe and Si, which reduced the effective amount of Fe available for formation of β and π phases, thereby reducing the size and volume of Fe-containing intermetallics, which, in turn, reduced the deleterious effect of Fe and improved the alloy strength and ductility.

Wan et al. [58] found that up to1.0 wt%, MM addition refined the grain size of cast Al-Cu-Mg-Si alloy and changed the eutectic Si morphology from needle-like and laminar to a granular type. Also, with the increase of the MM level, the tensile strength and elongation of the alloy first increased and then began to decrease. Alloy with 0.7 wt.% MM exhibited the highest mechanical properties. In another study, Chong et al. [59] studied the combined effects of P and MM on the microstructure and mechanical properties of hypereutectic Al-20%Si alloy. It was observed that, in general, alloys with the addition of 0.08% P and 0.6% MM exhibited highest mechanical properties and had the optimal microstructure compared to the alloy with no addition; refinement of the primary Si particles from 66.4 μm to 23.3 μm, and the eutectic silicon from 8.3 μm to 5.2 μm, was also noted. With respect to the mechanical properties, the ultimate tensile strength improved from 256 MPa to 306 MPa, and the ductility increased from 0.35% to 0.48%. **Figure** 7 shows the average primary Si size of the tested alloys with different P contents.

According to El Sebaie et al. [60–64] the presence of MM in unmodified and Sr-modified A319.1, A356.2 and A413.1 Al-Si casting alloys led to the following observations:

i. In general, the hardness values of the as-cast alloys were higher at high cooling rates than at low cooling rates. With MM, the hardness decreased at both solidification rates. **Figure 8** shows the hardness values obtained for these alloys after different aging treatments.

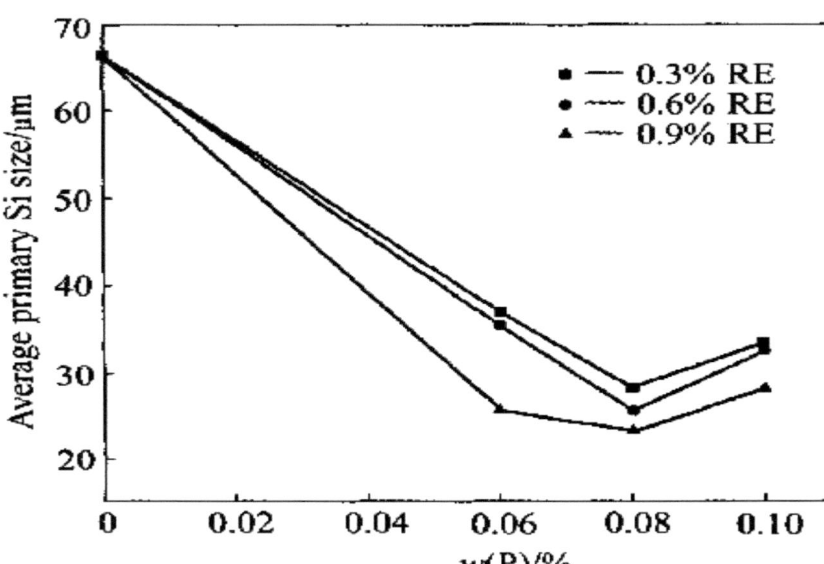

Figure 7.
Effect of P content on primary Si size of Al-20%Si alloys under same RE content condition [59].

ii. In the case of non-modified alloys MM addition resulted in partially modified eutectic silicon particles. This effect was more pronounced in the A413.1 and A319.1 alloys, compared to A356.2 alloy.

iii. The effect of MM as a modifier is more effective at high cooling rate (corresponding to DAS ~ 40 μm) than at the low cooling rate (DAS ~ 120 μm) for all the as-cast non-modified alloys.

iv. MM-containing intermetallic phases were observed at high and low cooling rates, each exhibiting a specific Ce/La ratio and morphology. Many of these MM-containing intermetallic phases were found to contain Sr., which confirmed the interaction of MM with Sr. – see **Figure 9** and Appendix A.

Combined addition of MM and Mn is an effective way to improve the strength of A390 alloy at elevated temperature by 25% [65]. Other studies by Zhu et al. [66, 67] reported the effects of 0.1–1.0 wt.% Ce-based MM additions on the microstructure, tensile properties and fracture behavior of as-cast and T6-treated A356 alloys. The main findings from their work are listed below for the specified conditions.

1.1.1 MM-modified as-cast A356 alloys

MM-containing intermetallic compounds cannot act as potential nucleate sites for the primary α-Al phase.

i. The modification effect of MM depends on the addition level. Minor additions of MM (less than 0.2 wt.%) result in partial modification, while more than 0.3 wt.% MM leads to full modification.

ii. The fracture path goes through the interdendritic region composed of eutectic silicon and MM-containing intermetallic compounds.

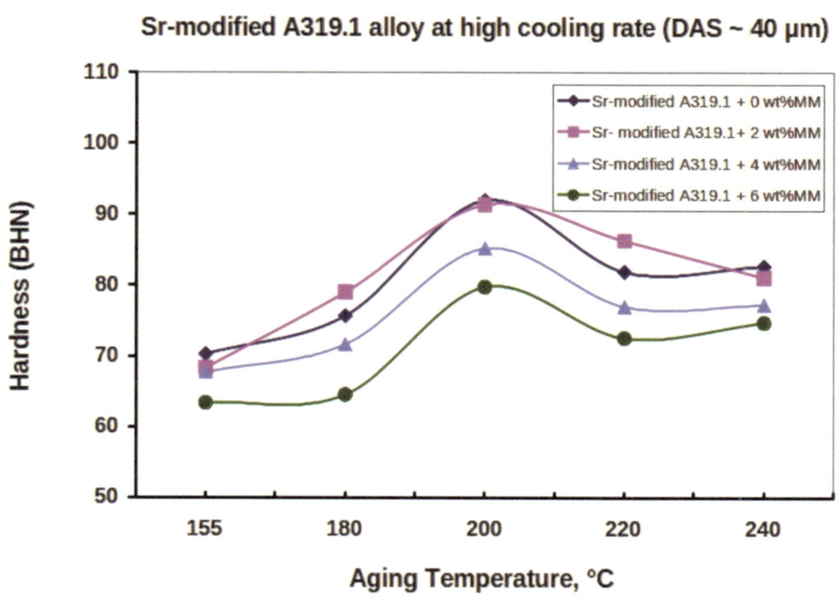

Figure 8.
Effect of mischmetal additions and aging temperature on the hardness of A319.1 alloys solidified at high solidification rates (DAS ~ 40 μm of Sr-modified conditions [63].

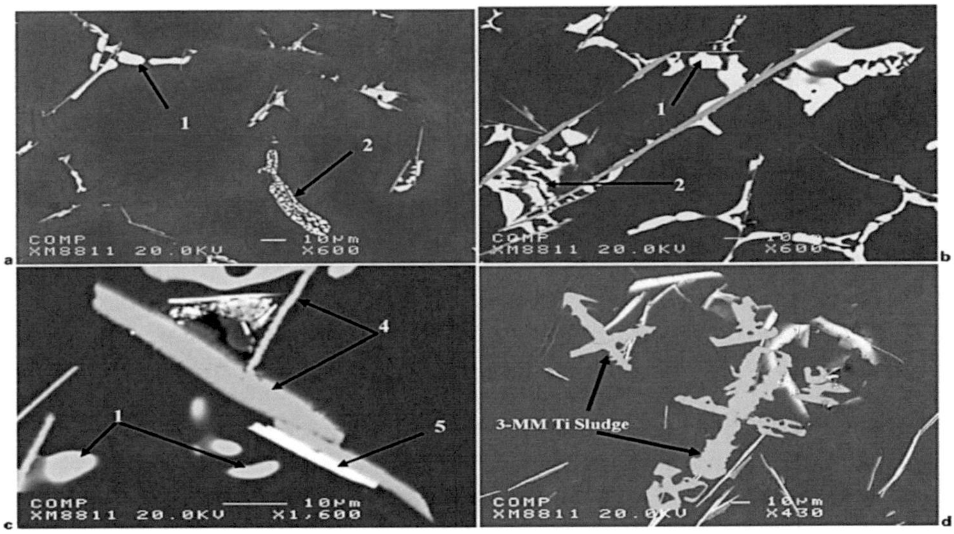

Figure 9.
Backscattered images of A319.1 alloy samples containing (a, b) 0 wt-% and (c, d) ***6*** *wt-%MM depicting intermetallic phases observed under high cooling rate conditions (S: Sr. modified; T: Heat treated samples) [64].*

1.1.2 MM-modified T6-A356 alloys

T6 heat treatment has great influence on the spheroidization of eutectic silicon particles in MM-modified A356 alloys than that in the unmodified alloys.

i. The UTS, YS, and %EL of the T6-treated A356 alloys with and without modification are improved due to the spheroidization of eutectic silicon particles and Mg_2Si precipitation hardening.

ii. SEM images show that the MM-modified T6 A356 alloy undergoes ductile fracture. It is worth noting that the eutectic silicon and MM-containing intermetallic particles provide the weak locations for initiation of the fracture as displayed in **Figure 10**.

The effects of different addition levels (0.0–1.0 wt.%) of La-based MM and heat treatment on the microstructure and tensile properties of two different sections of Al-Si casting alloy A357 were studied by Mousavi et al. [68–70]. Optimum recommended levels of MM are 0.1 wt.% and 0.3 wt.% for thin and thick sections of the casting, respectively. Examination of the microstructure at high level of MM (0.5 wt.%) exhibited the precipitation of a new AlSiLa intermetallic phase as shown in **Figure 11**.

The results of Jiang et al. [71] on the microstructure, tensile properties and fracture behavior of as-cast and T6 A357 alloy revealed that addition of MM reduced the size of the primary α-Al dendrites i.e., the SDAS value, and also improved the eutectic Si particle morphology. Accordingly, the MM-modified A357 alloy exhibited improvement in the tensile properties, particularly the elongation, in the T6-treated condition. The fracture surface of the tensile-tested sample of the unmodified alloy showed a clear brittle fracture, whereas that of the MM-modified A357 alloy exhibited dimple rupture and cracked eutectic Si particles, resulting in superior ductility. The results of Zhang et al. [72, 73] demonstrated that the AlTiB-MM addition to A356 alloy provided the most effective and synergetic grain size refinement compared to individual AlTiB or MM additions. Also, the properties of

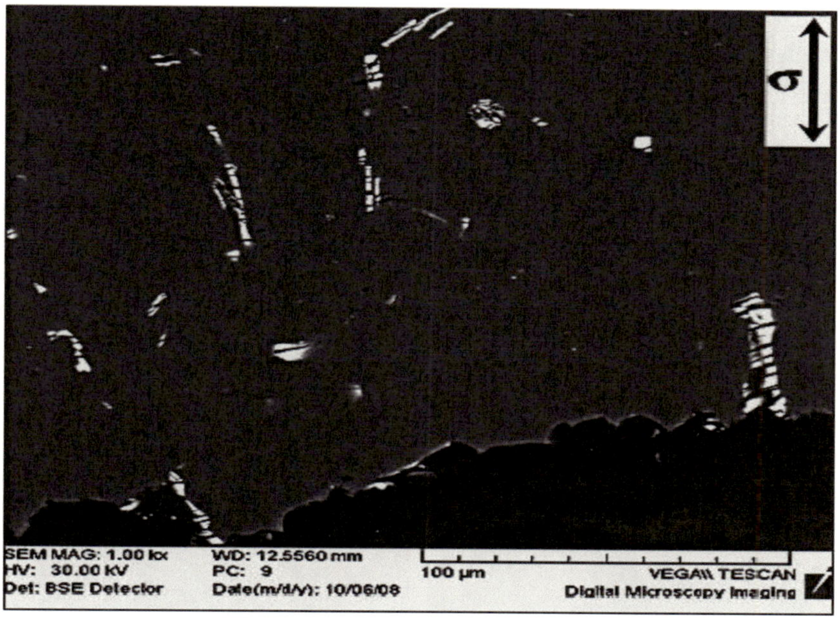

Figure 10.
Fracture surface parallel to the tensile direction of the T6-A356 alloys modified by 0.5 wt.% MM showing intercrystalline crack of RE-containing intermetallic compounds at the fracture surface [67].

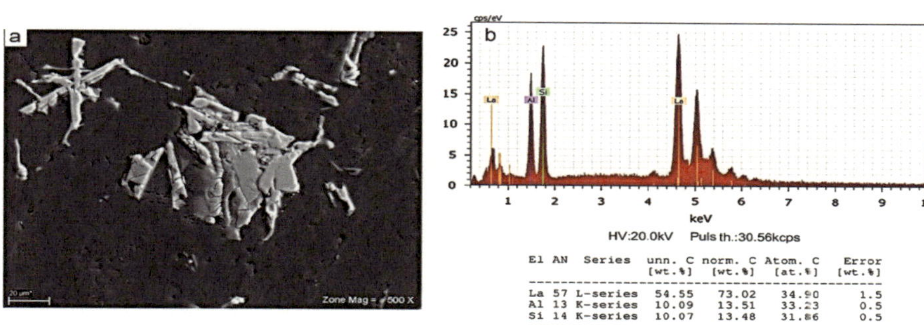

Figure 11.
(a) SEM photograph of Al–Si–La compound intermetallic in A357 alloy with 0.5 wt% mischmetal casting in thick section mold and (b) EDS spectrum showing the distribution of Al, Si and La in the intermetallic [70].

A356 alloy wheel refined by the AlTiB-MM were improved significantly. The tensile strength, yield strength, and elongation of the wheel spokes improved by approximately 11.3%, 10.8% and 44.1%, respectively.

Dang et al. [74] investigated the effects of the use of rare earth (RE) in the form of Al-10% RE master alloy (a mixture of Ce and La) and pouring temperature (1124 K through 1524 K in increments of 100 K) on the microstructure and mechanical properties of T6-treated Al-25%Si alloy. The authors observed that for the unmodified alloy, the primary Si morphology was transformed from platelets to fine polyhedral form, and the average size decreased with increase in pouring temperature (from 125 μm at 1124 K to ~62 μm at 1524 K). With a 1.2 wt.% RE addition, the primary Si exhibited a small blocky morphology, with an average particle size of 47 μm. In addition, the study showed that T6 MM-Al-25%Si alloy exhibited an improvement in the mechanical properties compared to the unmodified alloy, where the maximum tensile strength and elongation (208.3 MPa and 1.01%) were obtained for the sample modified with 1.2 wt.% RE followed by the T6 heat treatment. Tensile fracture exhibited three stages: microcrack initiation, crack coalescence, and quick crack propagation as shown in **Figure 12**.

According to Mahmoud et al. [75, 76], depending upon the amount of added Ti, two RE-based intermetallics can be formed: (i) a white phase, mainly platelet-like (approximately 2.5 μm thick), that is rich in RE, Si, Cu, and Al, and (ii) a second phase made up of mainly gray sludge particles (star-like) branching in different directions. The gray phase is rich in Ti with some RE (almost 20% of that in the white phase) with traces of Si and Cu. There is a strong interaction between RE and Sr., leading to a reduction in the efficiency of Sr. as a eutectic Si modifier, causing particle demodification. **Figure 13(a)** shows the actual morphology of the white phase which is likely to be thin platelets about 2.5 μm thick. The morphology of the gray sludge is well illustrated in **Figure 13(b)** exhibiting the branching of the gray phase particles in different directions.

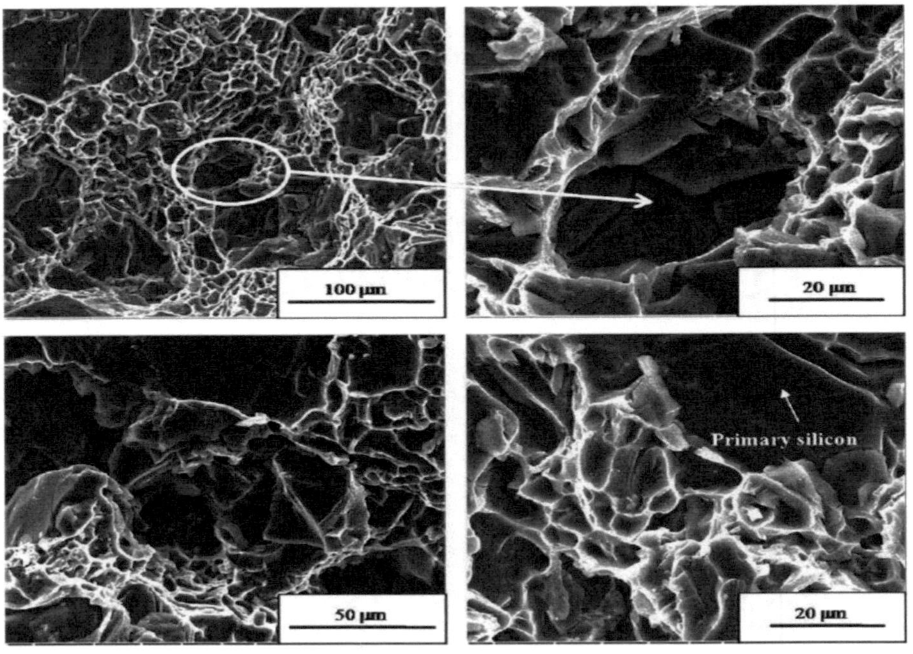

Figure 12.
The fracture morphology of Al-25% Si alloy in: (a) and (b) 1.2% RE, (c) and (d) 1.2% RE + T6 heat treatment [74].

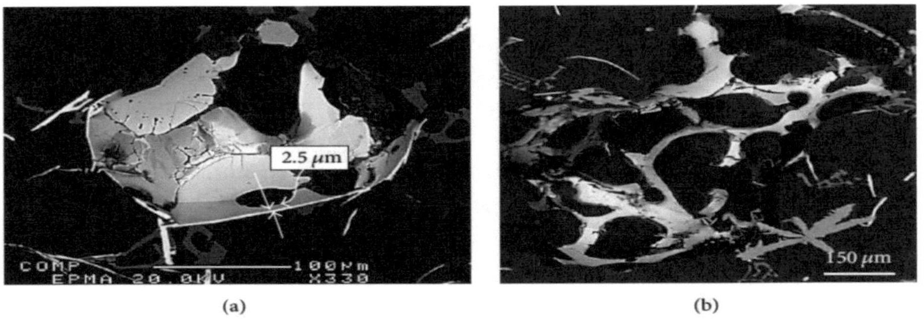

Figure 13.
RE-based intermetallic phases: (a) platelet-like phase; the two straight arrows pointing towards each other simply highlight the thickness of the platelet-like phase and the curved arrow that links the label 2.5 μm to the platelet indicates the actual thickness, (b) the gray sludge phase [76].

1.2 Effects of cerium additions

The results arising from the investigations of Song et al. [77, 78] showed that individual addition of 0.3%Ce or 0.2%Ti to Al-Cu-Mg-Ag alloys can decrease the grain size of the as-cast alloy, increase the nucleation rate of the Ω (metastable Al_2Cu) phase, inhibit the growth of the Ω phase during aging, and thereby increase the volume fraction and decrease the spacing of the Ω phase. Based on these microstructural observations, the yield strength and tensile strength of the alloy are increased. However, combined addition of Ce and Ti led to the formation of (Ce, Ti)-containing intermetallic compounds and increased the grain size during casting, with no influence on the nucleation and growth of the Ω phase during aging. The alloy containing both Ce and Ti had a relatively lower Vickers hardness and strength compared to the alloy containing individual additions of Ce or Ti. In another research, Song et al. [79] reported that Ce improved the thermal stability of the Ω phase by decreasing the diffusion velocity of the Cu atoms, and hence decreasing the coarsening speed of the phase, as well as through the aggregation of Ce atoms at the Ω phase/Al matrix interface, increasing the energy barrier for the thickening of the Ω phase plates which coarsen through a ledge nucleation mechanism. The strength of the Al-Cu-Mg-Ag alloy is improved, as a result.

The results on the effects of different levels of Ce additions on the microstructure, thermal behavior and mechanical properties of hypereutectic AlSi17CuMg alloy illustrated that addition of Ce (up to 1.0 wt.% Ce) can achieve refinement of the primary and eutectic silicon morphology. In general, alloy containing 1.0 wt.% Ce exhibited the best results with respect to the microstructural and strength properties. It was also observed that with 1.0 wt.% Ce, the alloy produced the highest reduction in the liquidus temperature from 686.6 °C to 591.9 °C. [79].

The eutectic silicon modification in A356 alloy modified with 1.0 wt.% Ce was greatly improved [80]. However, the thermal analysis revealed that there is no direct relation between the eutectic growth temperature and silicon modification. The microstructural characterization showed that two kinds of Ce-containing intermetallic phases were found, including Ce-23Al-22Si and Al-17Ce-12Ti-2Si-2 Mg (in wt.%). While the ductility of the Ce-modified alloys was enhanced for Ce additions of 0.6 wt.% and above, there was no positive effect on the ultimate tensile strength; this was attributed to the formation of the Al-17Ce-12Ti-2Si-2 Mg phase which reduced the amount of free Mg available for precipitation of the Mg_2Si strengthening phase.

The effects of various concentrations of Ce (0.0, 0.01, 0.02, 0.05 and 0.1 wt.%) on the solidification and mechanical properties of AA A360 (Al-10%Si-0.5%Mg) alloy were reported by Voncina et al. [81]. The results showed that the solidus temperature decreased with increasing Ce addition. The eutectic (α_{Al} + Mg_2Si) temperature also decreased with Ce addition. It was found that the precipitation enthalpy decreased with the Ce addition, while precipitation occurred more rapidly and intensively, indicating increased reaction kinetics. Mechanical properties like tensile strength and hardness also increased with the Ce addition, where the hardness of the investigated alloy could be attributed to the phase composed of Al, Ce, Mg and Si. The precipitation enthalpy also decreased with increasing Ce addition and increased with increasing cooling rate as determined from simple DSC analysis, as shown in **Figure 14**. It was anticipated that Ce in A360 alloy decreases the activation energy for the precipitation of the Mg_2Si phase and consequently precipitation enthalpy.

It is worth noting that in another study Voncina et al. [82] reported that addition of Ce to A380 alloy led to a change in the morphology of eutectic Al_2Cu phase from "crumbled" to fully formed (finer eutectic-like to blocky form) and caused the

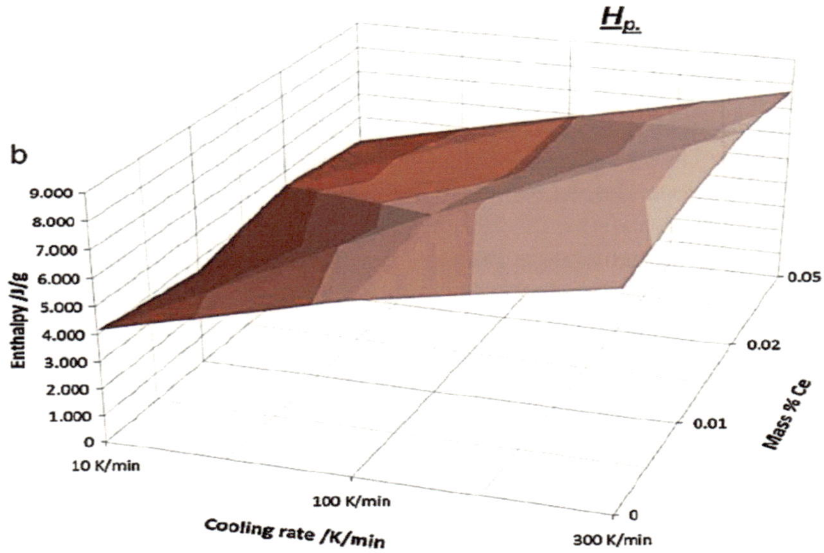

Figure 14.
Precipitation enthalpy regarding the Ce addition and cooling rate at heating DSC curves.

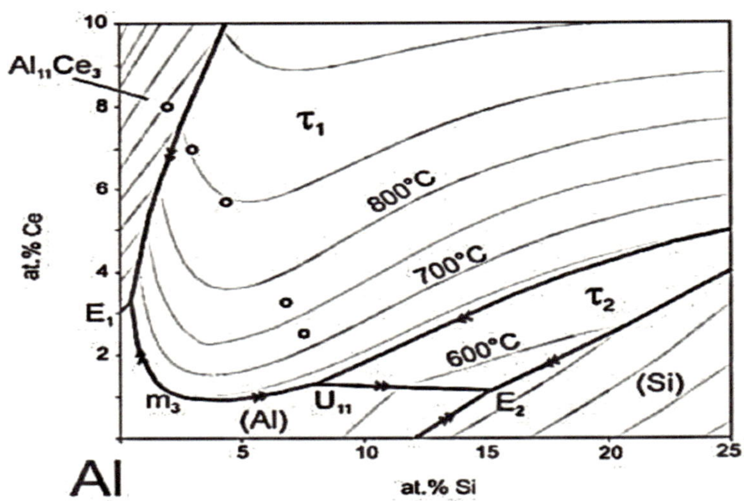

Figure 15.
Al-corner of ternary Al-Si- Ce system [82].

formation of small primary crystals of α_{Al} which resulted in grain refining of the alloy. A needle-shaped AlCeCuSi phase ($Al_9Ce_2Cu_5Si_3$) was also detected. The solidification of hypoeutectic Al-Si alloys with Ce addition can be described by the Al-corner of ternary system Al-Si-Ce (**Figure 15**).

Effects of Ce-Sr interaction on the nucleation of primary α-Al phase dendrites in hypoeutectic Al-7%Si-Mg cast alloy were examined by Chen et al. [83]. It was found that with addition of Ce and Sr., the grain size of the dendritic α-Al phase becomes well refined, decreasing from 150 μm to 90 μm, and is attributed to the exponential increase in nucleation frequency (10^{24}), compared to the unmodified alloy, and restricted growth. Increasing the Ce level (0, 0.3, 0.5, 0.8 and 1 wt.%) in Al-20%Si alloy would cause a significant refinement of the primary Si crystals with the change in the morphology of the eutectic Si phase from coarse platelet like to a fine fibrous structure [84]. Accordingly, the ultimate tensile strength (UTS) and elongation (% El) increased by 68.2% and 53.1%, respectively, as a result of these effects.

Ye et al. [85] investigated the influence of Ce content (0.2% and 0.4 wt.%) on the impact properties and microstructures of 2519A aluminum alloy, a new version of 2519 alloy (with a higher Cu/Mg ratio) used for armor material. Based on the results of their research, it was found that 0.2 wt.% Ce addition leads to an increase in the volume fraction of the precipitation phase, in addition to a more dispersive and homogeneous distribution of finer θ' (Al_2Cu) precipitates, which result in improving the ability of the alloy for absorbing impact energy. Formation of Al_8Cu_4Ce phase which is thermally stable at high temperature is also expected to enhance the high temperature mechanical properties of the alloy. Yii et al. [86] reported that the addition of Ce in the Al-20%Si alloys refined the Si primary phase as the Ce additions were increased. The results also showed that addition of Ce in the range of 0.46 to 2.24 wt.% led to the formation of fine cells dispersed in the Al-matrix. These cells consisted of a mixture of eutectic Si particles and Ce-containing intermetallic phases (Al_3Ce and $CeAl_{1.2}$ $Si_{10.8}$). The amount of the intermetallic phases increased with increasing Ce addition.

Promising scientific investigations were made by Ahmad and coworkers [87, 88] on the influence of Ce on the microstructure of a commercial Al-11%Si-Cu-Mg eutectic cast alloy (ADC12). The main findings from their studies are summarized below.

i. The addition of Ce to ADC12 alloy leads to improvement in the Si particles modification and reduces the Si particle size by 62% [87].

ii. Cooling rate has no significant effect in the 1.5 wt.% Ce-modified ADC12 alloy, compared to the base alloy and this may be attributed to the formation of intermetallic phases.

iii. Investigation of the Al-Si eutectic phase using the thermal analysis technique showed that addition of Ce had a significant effect on the nucleation, growth, and minimum temperatures of Al-Si, and decreased as the Ce concentration increased; refinement of the Si structure was observed up to 1.0 wt.% Ce. In addition, the growth and nucleation temperatures of the Al–Cu phase, which is the last phase to solidify, also increased with increasing level of Ce. The formation of Ce-containing intermetallic compounds such as Al-Si-Ce and Al-Si-Cu-Ce affected the degree of Si modification [88] - **Figure 16** and **Table 4**.

iv. Ce addition refined the secondary dendrite arm spacing (SDAS) by approximately 36%. In addition, the tensile strength and quality index of Al-11%Si-Cu-Mg increased to 237.6 MPa and 265 MPa, respectively, after the addition of 0.1 wt.% Ce.

Effect of solidification rate and rare earth metal addition on the microstructural characteristics and porosity formation in A356 alloy was investigated by Mahmoud et al. [89]. According to the atomic radius ratio, γLa/γSi is 1.604 and γCe/γSi is 1.559, theoretically, which shows that Ce is relatively more effective than La. These findings confirm that Sr. is the most dominating modification agent. Interaction between rare earth (RE) metals and Sr. would reduce the effectiveness of Sr. Although modification with Sr. causes the formation of shrinkage porosity, it also reacts with RE-rich intermetallics, resulting in their fragmentation. **Figure 17** reveals the distribution of La, Ce, and Sr. in RE-rich platelets, which explains the

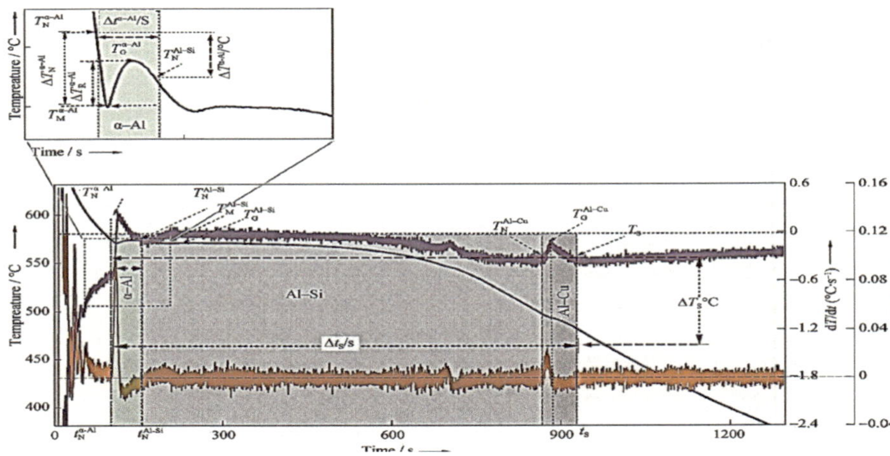

Figure 16.
Cooling curve and the first and second derivative of the 0.1 wt% Ce-containing alloy with characteristic parameters and α-Al phase arrest regions, showing points of interest [88].

Parameter	Description
$T_N^{\alpha-A1}/^\circ C$	Nucleation temperature of α-A1 (onset of a-A1 phase)
$T_M^{\alpha-A1}/^\circ C$	Minimum temperature of α-A1
$T_G^{\alpha-A1}/^\circ C$	Maximum growth temperature of α-A1
$T_R^{\alpha-A1}/^\circ C$	Recalescence temperature = $T_G^{\alpha-A1} - T_M^{\alpha-A1}$
$T_N^{\alpha-A1}/^\circ C$	Nucleation undercooling temperature = $T_N^{\alpha-A1} - T_M^{\alpha-A1}$
$\Delta T^{\alpha-A1}/^\circ C$	Solidification temperature range of α-A1 phase = $T_N^{\alpha-A1} - T_N^{A1-Si}$
$T_N^{A1-Si}/^\circ C$	Nucleation temperature of Al-Si (end of α-A1 phase)
$T_s/^\circ C$	Solidus temperature (end of solidification)
t_s/s	Solidus time (end of solidification)
$\Delta T_s/^\circ C$	Total temperature range = $T_N^{\alpha-A1} - T_s$
$\Delta t_s/s$	Total temperature range = $t_N^{\alpha-A1} - t_s$

Table 4.
Solidification characteristic parameters identified during solidification of the α-A1 phase and at the end of solidification.

partial modification of the surrounding Si particles as less Sr. is available for modification of the eutectic Si.

1.3 Effects of lanthanum additions

It is inferred from the work of Mahmoud et al. [90] using high purity (99.5%) lanthanum or cerium, that La and Ce have more or less the same effect on the microstructures, with the La- and Ce-containing intermetallics displaying similar morphologies. Regardless of the alloy composition, an addition of 150 ppm Sr. or 0.2% RE results in improving the UTS by 25–52% in the T6 condition, with a decrease in ductility from 3% to 2.1%. The addition or RE metals (La + Ce) up to 3 wt.% leads to an increase in the freezing range through an increase in the melting point of the non-modified alloys, with decrease in the Al-Si eutectic temperature, by 12 °C and 8 °C, respectively, at 3 wt.% addition, **Figure 18**. The authors

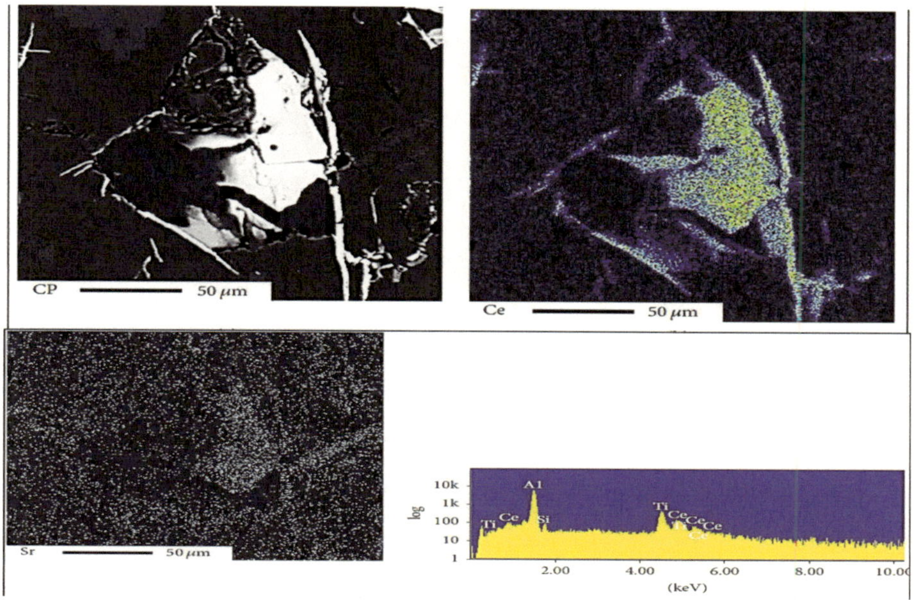

Figure 17.
(a) Backscattered electron image showing Ce-rich platelets in Sr-modified A356 alloy containing 1.025% wt.% Ce, and elemental distribution of (b) Ce, (c) Sr., and (d) EDS spectrum corresponding to (a) [89].

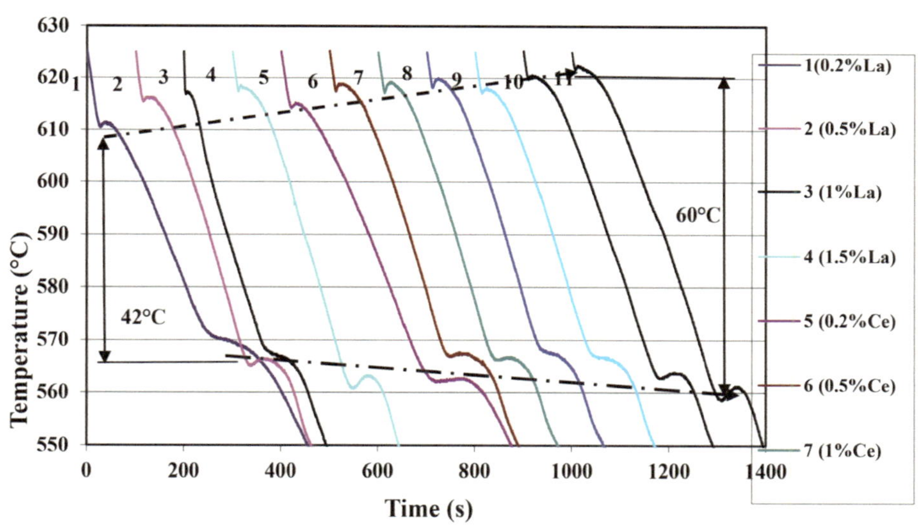

Figure 18.
Solidification curves of unmodified 356 alloy [90].

concluded that the addition of La or Ce leads only to fragmentation of the Si platelets in the case of non-modified alloys and only partial modification in the case of Sr-modified alloys. The direct advantage of the addition of RE metals to non-grained alloys is the reduction in the grain size by about 50% at 3 wt.% RE addition.

According to Song et al. [79], when RE was added to 356 alloy in the amount of 0.6 wt.%, the mean grain size was reduced by about 50%. A similar effect was observed in the work of Ibrahim et al. [91, 92] on the effect of rare earth metals on the mechanical properties of 356 and 413 alloys, as shown in **Figure 19**, in particular, in alloys 3 L and 4 L, see **Figure 19** (b, e). Due to Ce-Ti interaction, the grain refining effectiveness was reduced in the 3C and 4C alloys, as shown in **Figure 19**

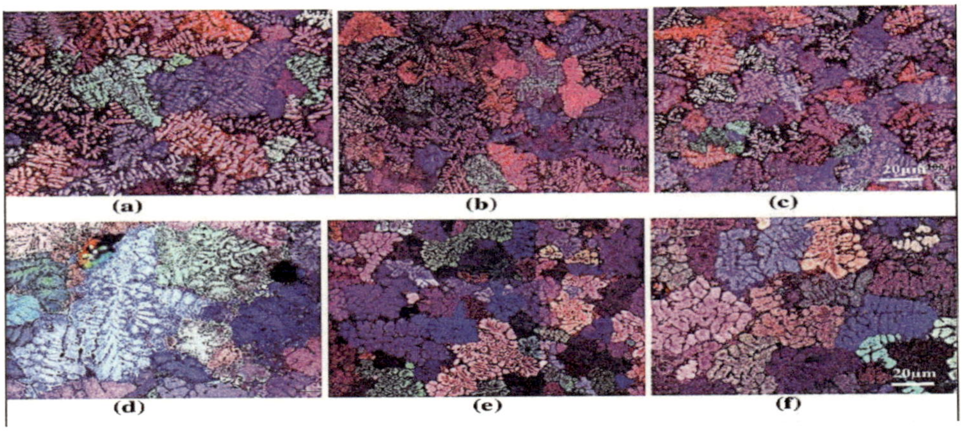

Figure 19.
Effect of La and Ce addition on the grain size in A356 alloys: (a) alloy 3, no RE addition, (b) alloy 3 L, addition of 1% La, (c) alloy 3C, addition of 1% Ce. Effect of La and Ce addition on the grain size in A413 alloys: (d) alloy 4, no RE addition, (e) alloy 4 L, addition of 1% La, (f) alloy 4C, addition of 1% Ce.

(c, f). Microstructural characterization of Al-Si cast alloys containing rare earth additions was performed by Elgallad et al. [93]. The main findings of this study were that the addition of La and/or Ce resulted in the formation of a whitecoloured Al-Si-La/Ce/(La,Ce) phase in both A356 and A413 alloys. In addition, the presence of Ti in the A356 alloy allowed for the formation of a gray-colored Al-Ti-La/Ce phase besides the Al-Si-La/Ce/(La,Ce) phase. The formation of these phases significantly increased the phase volume fraction of intermetallics in the A356 and A413 alloys. In the presence of Sr., the white-colored Al-Si-La/Ce/(La,Ce) phase was found to also contain Sr. (~1 at%). No specific Sr-La/Ce intermetallic phases were detected in the microstructures of the alloys investigated. **Figure 20** (i) shows the DSC cooling curves of the A413 alloy before and after the addition of La and Ce, individually or in combination, whereas **Figure 20** (ii) shows the DSC heating curves of the A413 alloy without and with La and/or Ce.

The results on various La additions on the microstructure of as-cast ADC12 (Al-11%Si-Cu-Mg) alloy [94] indicated that the α-Al and eutectic Si crystals were modified with the addition of 0.3 wt.% La. The eutectic Si crystals showed a granular distribution. At the same time, the alloy possessed the best mechanical properties. However, as the La addition was increased beyond 0.3 wt.%, the microstructure coarsened gradually and the mechanical properties decreased as a result.

Song et al. [77–79] analyzed the impact of different additions of La (0.0, 0.3, 0.6, and 0.9 wt.%) on the microstructure and hot crack resistance of ADC12 alloy. The results showed that, as the La added increased from 0.0% to 0.6 wt%, the structure of the α-Al phase gradually varied from a well-developed dendritic crystal into fine dendritic crystal, equiaxed crystal and spheroidal crystal; the eutectic silicon morphology varied from needle-like or tabular shape into a fine rod-like shape; the hot cracking force of the alloy also gradually decreased. Optimum alloy modification, alloy refinement and hot cracking resistance were achieved at 0.6 wt.% La addition. However, when the addition of La reached 0.9 wt.%, the excessive amount of La segregated at the grain boundaries, forming intermetallic phases.

Similarly, Chen et al. [84] evaluated the effects of combined addition of lanthanum and boron (B) on the grain refinement of Al-Si casting alloys, and found that such additions can effectively refine the grains of Al-Si alloys compared to individual addition of boron. This work also reported that with addition of La, the tensile properties of the alloy, in particular, the elongation are enhanced. The response of

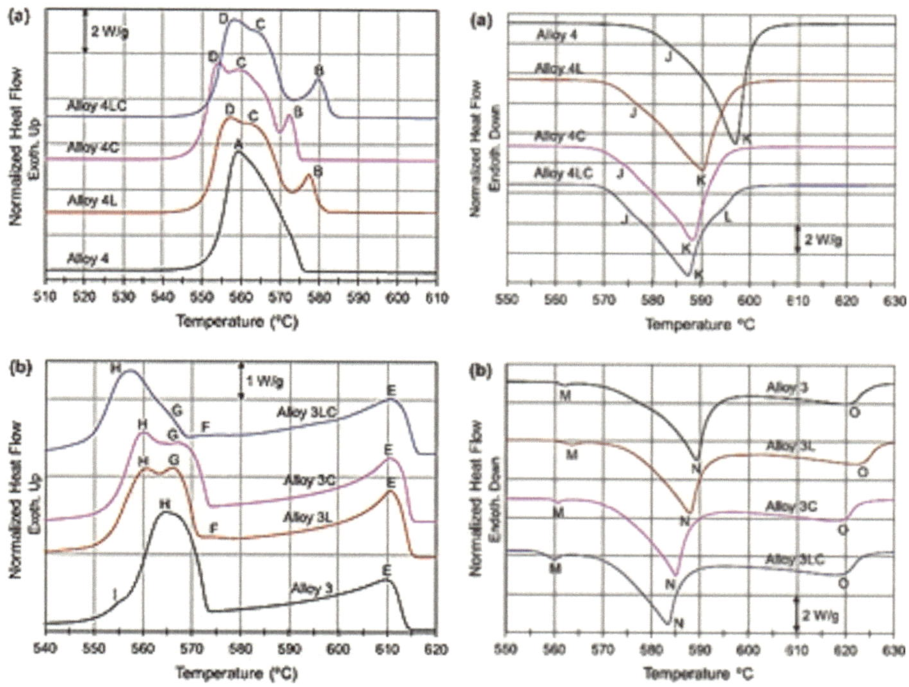

Figure 20.
(i) DSC cooling curves of (a) A413 and (b) A356 alloys, respectively, without and with La and/or Ce.(ii) DSC heating curves of: (a) A413 (alloy 4) and (b) A356 (alloy 3) alloys without and with La and/or Ce [93].

trace additions of La (0.05% - 0.1 wt.%) on the microstructures and tensile properties of B-refined and Sr-modified Al-11%Si-1.5%Cu-0.3%Mg casting alloys were investigated by Lu et al. [95] who found that introducing La/B in the weight ratio of 2:1 produced well refined α-Al grains and modified eutectic Si particles in the alloy, as well as strengthening intermetallic precipitates, which improved the ultimate tensile strength from 234 to 270 MPa, and elongation and from 4.0 to 5.8%, respectively.

A study on the synergistic effect of Sr. and La on the microstructure and mechanical properties of A356.2 alloy was carried out by Qiu et al. [96]. It was found that, with the addition of 0.5 wt.% Al-6Sr-7La master alloy, the alloy exhibited optimal microstructure and mechanical properties, with the secondary dendrite arm spacing (SDAS) decreasing to 17.9 μm and the acicular eutectic silicon transforming to a fibrous form. With the improved microstructure, the ultimate tensile strength, yield strength and elongation of the alloy (with 0.5 wt.% Al-6Sr-7La) increased to 228.15 MPa, 108.13 MPa and 11.92%, respectively, which were much better than those of the "traditionally treated" A356.2 alloy (using 0.2 wt.% Al-5Ti-1B for grain refining and 0.2 wt.% Al-10Sr for Si modification).

Tang et al. [97] investigated the effect of Sr. and La addition on the microstructure and mechanical properties of a secondary Al-Si-Cu-Fe alloy. The quantitative metallographic results indicated that addition of different levels of Sr. and La modification agents, added in the form of Al-10%Sr. and Al-10%La, produced varied refinement effects on the mean length of needle-like phases and the SDAS value. The total dosage of Sr. and La varied from 0.04 to 0.2 wt.% (Sr/La = 1:1). The minimum mean length of needle-like phases (Sr/La = 1:1) and the SDAS value (Sr/La = 1:5) were obtained by setting the addition amount of the modification agent at 0.12 wt.%. The mean length of the needle-like phase dropped from 364 to 55.3 μm, while the SDAS decreased from ~22 to 9.7 μm, i.e., by 84.5% and 55.8%, respectively.

Effect of solution treatment on the microstructure and mechanical properties of A356.2 alloy treated with Al–Sr–La master alloy was examined by Ding et al. [98] who found that the optimal solution treatment parameters for A356.2 aluminum alloy treated by Al–6Sr–7La are: solution treatment at 540 °C for 3 h, followed by quenching in 60 °C water – see **Figure 21**. The alloy under this condition possesses the optimal comprehensive conditions/values of microstructure, eutectic silicon morphology, UTS, YS, and EL, which is beneficial to the subsequent aging process.

The effects of Ti - La interaction on the microstructure and mechanical properties of B-refined and Sr-modified Al-11%Si alloys showed that the addition of 0.05 wt.% B induces a transformation of the eutectic Si from finely fibrous to coarse plate-like morphology in the Al-11%Si alloy modified with 0.02 wt% Sr., owing to the poisoning of impurity induced twinning (IIT) mechanism [99]. Thus, the eutectic Si growth occurred only by the twin plane re-entrant (TPRE) mechanism. Both Ti and La can neutralize the poisoning effect of the interaction between Sr. and B in the Al–11%Si alloy; however, the neutralizing effect of La is dependent on the addition sequence. The combined addition of La and B elements promoted the effective refinement of α-Al grains, but an inhomogeneous modification of the eutectic Si phase was also observed, leading to a slight decrease in the elongation. The poisoning effect can also be proved by the reduction of multiple Si twins as shown in **Figure 22**. **Figure 23** display the affinity of RE metals to react with Sr. leading to the observed loss of modification in the present alloys [100].

(a) 20 °C; (b) 30 °C; (c) 40 °C; (d) 50 °C; (e) 60 °C; and (f) 70 °C [98].

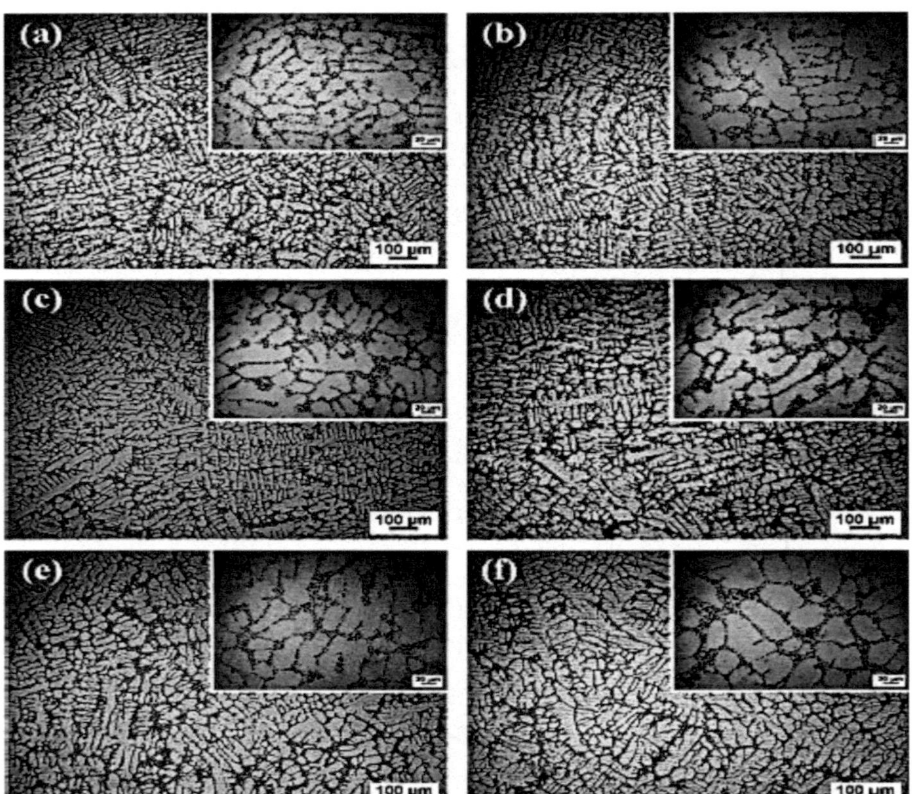

Figure 21.
Effect of quenching temperatures on the microstructure of A56.2-Sr-La alloy: (a) 20 °C; (b) 30 °C; (c) 40 °C; (d) 50 °C; (e) 60 °C; and (f) 70 °C [98].

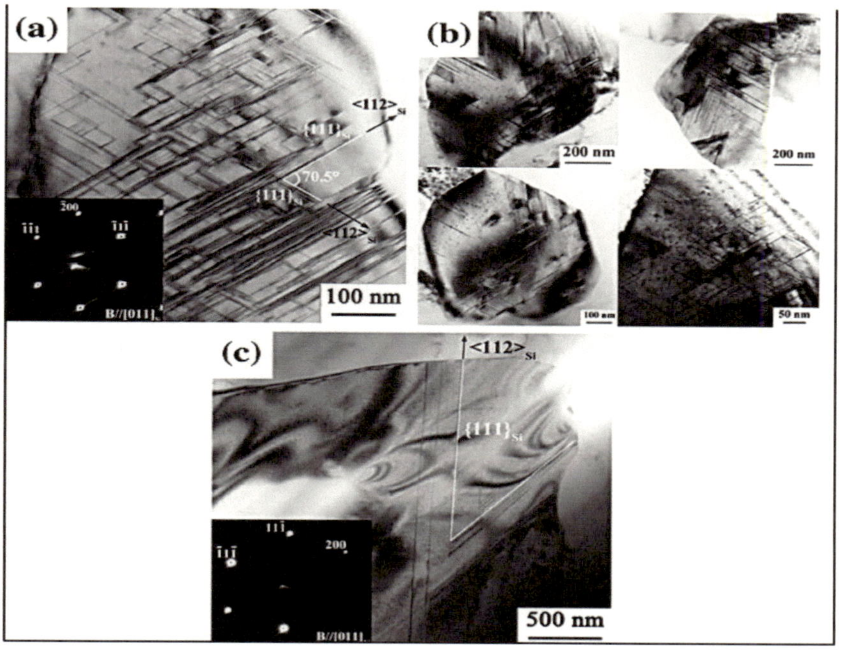

Figure 22.
(a) TEM bright field image and corresponding selected area diffraction pattern of Si particle, tilted to [011]Si zone axis, in the S5 alloy (0.1216% La, 0.0526%B, 0.0218%Sr); (b) assembly of TEM bright field images of different Si particles taken from the S5 alloy; (c) TEM bright field image and corresponding selected area diffraction pattern of Si plates, tilted to [011]Si zone axis, in the S6 alloy.

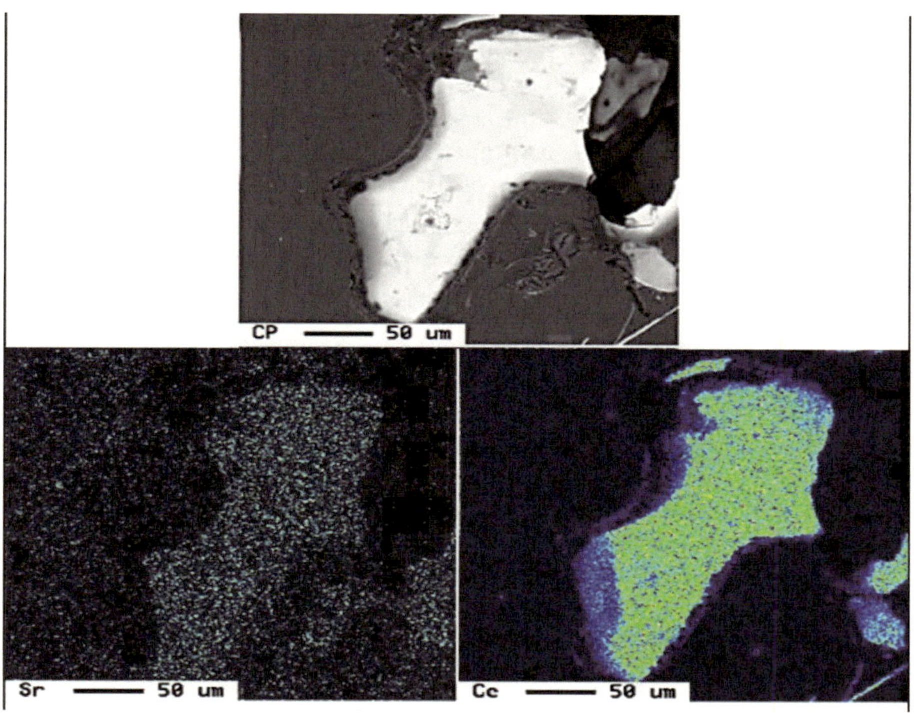

Figure 23.
Ce-Sr interactions in A356 alloy modified with 1.0% Ce + 0.01% Sr. [100].

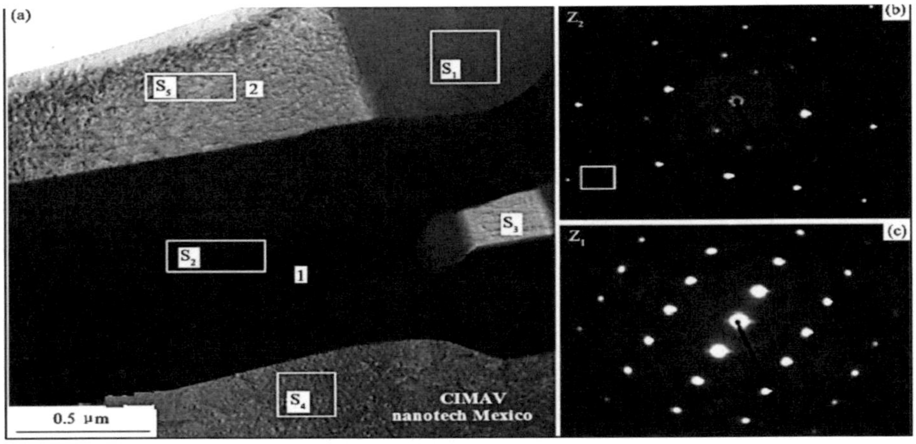

Figure 24.
Bright field TEM micrograph of the La/Ce, Si and Al phases (a) squares indicate where elemental analysis was carried out; (b, c) isolated numbers indicate the zone where diffraction patterns were taken.

1.4 Effects of mixed cerium and lanthanum (RE) additions

The microstructure and mechanical properties of the automotive A356 aluminum alloy reinforced with 0.2 wt.% Al-6Ce-3La (coded ACL) were investigated by Aguirre-De la Torre et al. [101]. In this study, the ACL was added to the molten A356 alloy in the as-received condition and also processed by another route, employing mechanical milling and powder metallurgy techniques. Scanning electron microscopic observations indicated a homogeneous dispersion of La/Ce phases using both routes. In regard to the mechanical properties, however, the modified A356 alloy with the ACL added in the as-received condition, showed an improvement in the mechanical performance of the A356 alloy over that reinforced with the mechanically milled ACL. A bright field TEM micrograph from the FIB-milled TEM sample is presented in **Figure 24** revealing the presence of three phases in different shades of gray as observed in **Figure 24(a)**.

Also, Wang et al. [102] studied the effects of mixed La and Ce rare earth additions on the microstructure and properties of Al-0.75%Mg-0.6%Si alloy. The results showed that the mixed addition of La and Ce had a positive effect on the grain refinement of the investigated alloy. Accordingly, the tensile strength and elongation of Al-0.75%Mg-0.6%Si alloy gradually increased with the increase in the amount of La and Ce added.

Another study was carried out by Du et al. [103–105] on the influence of 0.25 wt. % and 0.50 wt.% mixed additions of Ce and La on the microstructure and mechanical properties of an Al-Cu-Mn-Mg-Fe alloy. With the mixed addition, two intermetallic phases, Al_8Cu_4Ce and Al_6Cu_6La, were formed. The results also showed that the 0.25 wt.% addition could promote the formation of a denser precipitation of $Al_{20}Cu_2Mn_3$ and $Al_6(Mn,Fe)$ phases, which improved the mechanical properties of the alloy at room temperature. However, up to 0.50 wt.% Ce-La addition promoted the formation of coarse Al_8Cu_4Ce phase, in addition to the Al_6Cu_6La and $Al_6(Mn, Fe)$ phases, which resulted in weakened mechanical properties [106, 107].

2. Summary

The review of the literature presented in this chapter has highlighted the numerous studies that have been carried out on the effects of rare earth elements, in particular, Ce and La, on the microstructure and mechanical properties of aluminum alloys. While a number of these investigations have been undertaken by Chinese researchers, due likely to the easily available natural source of rare earths in the form of mischmetal, studies by other researchers are also reported. Previous studies carried out by the TAMLA research group have investigated the influence of rare earth elements and mischmetal on the performance of A356, A413.1 and other Al-Si alloys. With the more recent focus on the development of new Al-Cu based alloys for high temperature performance of automotive components, it was considered worthwhile to also investigate the effects of Ce and La rare earth metal additions to these alloys, taking into consideration low and relatively high Si levels. Putting the importance of rare earth elements into proper perspective, recently, University of Kentucky researchers have reported producing nearly pure rare earth concentrates from Kentucky coal sources [103–105]. The patent pending process developed by Honaker and Zhang is a low cost and environmentally friendly recovery process. Interest in rare earth elements is currently at its peak in the U.S.A., with the Department of Energy investing millions in research, as REs constitute essential components of diverse technologies in the high-tech and renewable energy industries.

Appendix A- Mischmetals

Phase no.	Element	wt-% Av.	at-% Av.	Calculated formula	Shape and color	Suggested formula
1	Al	59.76	70.74	$Al_{16}(MnFe)_4Si_3$	Chinese script, medium gray	$Al_{15}(MnFe)_3Si_2$ + α. Iron
	Si	10.07	11.44			
	Fe	20.3	11.6			
	Mn	8.31	4.83			
	Total	98.44	98.61			
2	Al	23.21	41.32	$Al_{10}(CeLaPrNd)_4Si_9$ (Ce/La = 1.57:1)	Chinese script, white	$Al_2MM^*Si_2$†
	Si	21.99	38.77			
	Ce	26.93	9.47			
	La	17.05	6.03			
	Pr	6.83	2.31			
	Nd	2.66	0.9			
	Total	98.67	98.8			
3	Al	22.7	41.06	$Al_{10}(CeLaPrNd)_4Si_9$ (Ce/La = 1.48:1)	Chinese script, white	$Al_2MM^*Si_2$† + 0.48 wt-%Sr
	Si	22.5	39.11			
	Ce	26.65	9.28			
	La	17.78	6.24			
	Pr	6.53	2.26			
	Nd	2.39	0.8			
	Sr	0.48	0.26			
	Total	99.03	99.01			

Phase no.	Element	wt-% Av.	at-% Av.	Calculated formula	Shape and color	Suggested formula
4	Al	31.02	52.6	$Al_{10}(CeLaPrNd)_2(CuNi)_3Si_3$ (Ce/La = 2.1:1)	Plate. like, medium light gray	Al_5MM^* (CuNi) $Si^†$ + excess of Al
	Si	10.92	18.58			
	Cu	14.02	9.03			
	Ni	6.05	5.40			
	Ce	21.78	7.05			
	La	10.41	3.36			
	Pr	5.12	1.57			
	Nd	1.91	0.57			
	Total	101.2	98.16			

MM: mischmetal.
†*Al reading could be higher than the actual content due to the small size of the examined particles.*

Table A.
Chemical compositions of intermetallic phases observed in as-cast A413.1 alloy containing 6 wt.% MM (WDS analysis, SDAS:120 mm) [62].

Appendix – B Ce addition

Phase Color	Elements (At.%)						Suggested phase
	Al	Ti	Fe	Cu	Si	Ce	
Gray-1	84.41	6.426	0.025	2.050	1.526	4.009	$Al_{21}Ti_2Ce$ (with trace of Cu and Si)
Gray-2	83.08	6.516	0.052	2.308	2.726	4.053	$Al_{21}Ti_2Ce$ (with trace of Cu and Si)
Gray-3	83.88	6.696	0.015	1.963	2.278	4.019	$Al_{21}Ti_2Ce$ (with trace of Cu and Si)
Gray-4	83.66	6.423	0.013	2.278	2.331	3.972	$Al_{21}Ti_2Ce$ (with trace of Cu and Si)
Gray-5	84.60	6.630	0.025	1.734	1.872	4.003	$Al_{21}Ti_2Ce$ (with trace of Cu and Si)
White-1	40.54	0.000	0.069	12.38	26.96	18.58	$Al_9Ce_4Cu_2Si_4$
White-2	40.44	0.000	0.069	11.80	27.73	18.56	$Al_9Ce_4Cu_2Si_4$
White-3	46.08	0.000	0.112	12.04	24.28	16.52	$Al_9Ce_4Cu_2Si_4$
White-4	40.21	0.000	0.122	12.97	27.04	18.45	$Al_9Ce_4Cu_2Si_4$
White-5	40.57	0.000	0.096	12.24	26.96	18.78	$Al_9Ce_4Cu_2Si_4$

Table B-1.
WDS analysis of RE intermetallic phases observed with 1.0 wt.% Ce [70].

Phase Color	Elements (At.%)						Suggested phase
	Al	Ti	Fe	Cu	Si	Ce	
gray-1	85.78	6.314	0.014	1.080	1.388	4.156	$Al_{21}Ti_2La$(with trace of Cu and Si)
gray-2	85.40	6.228	0.015	1.410	1.431	4.134	$Al_{21}Ti_2La$ (with trace of Cu and Si)
gray-3	84.52	6.255	0.019	1.777	1.869	4.265	$Al_{21}Ti_2La$ (with trace of Cu and Si)

Phase Color	Elements (At.%)						Suggested phase
	Al	Ti	Fe	Cu	Si	Ce	
gray-4	83.99	6.394	0.008	2.558	1.601	4.187	$Al_{21}Ti_2La$ (with trace of Cu and Si)
gray-5	85.53	6.377	0.015	0.946	1.659	4.272	$Al_{21}Ti_2La$(with trace of Cu and Si)
white-1	51.26	0.000	0.000	0.073	25.64	21.82	Al_2CeSi
white-2	45.37	0.000	0.000	0.276	28.55	24.59	Al_2CeSi
white-3	45.49	0.000	0.017	0.292	28.42	24.56	Al_2CeSi
white-4	45.15	0.000	0.014	0.259	29.20	24.20	Al_2CeSi
white-5	51.86	0.000	0.009	0.077	25.33	21.46	Al_2CeSi

Table B-2.
WDS analysis of RE intermetallic phases observed with 5.0 wt.% Ce [75].

Author details

Mohamed Gamal Mahmoud[1], Yasser Zedan[2], Agnes-Marie Samuel[3], Victor Songmene[2], Herebert W. Doty[4] and Fawzy H. Samuel[3]*

1 Department of Mechanical Design and Production (MDP), Cairo University, Faculty of Engineering, Egypt

2 Département de Génie Mécanique École de Technologie Supérieure, Montréal (Qc), Canada

3 Département des Sciences Appliquées, Université du Québec à Chicoutimi, Québec, Canada

4 Materials Technology, General Motors Global Technology Center, Warren, MI, USA

*Address all correspondence to: fhsamuel@uqac.ca

References

[1] J. L. Murray A. J. McAlister, The Al-Si (Aluminum-Silicon) system, Bulletin of Alloy Phase Diagrams volume, 1984, 74, https://doi.org/10.1007/BF02868729.

[2] M. F. Ashby and D. R.H. Jones, in Engineering Materials 2 (Fourth Edition), 2013, Eutectic structure

[3] M. D. Sabatino and L. Arnberg, Castability of aluminium alloys, Transactions of The Indian Institute of Metals, 2009, 62, pp. 321–325

[4] G. Timelli and F. Bonollo, "Microstructure, Defects and Properties in Aluminum Alloys Castings: A Review," Proc. Int. Conf. Aluminium Two Thousand, Firenze (2007).

[5] Y.W. Lee, E. Chang, C.F. Chieu, "Modeling of Feeding Behavior of Solidifying Al-7Si-0.3Mg Alloy Plate Casting," Metall. Trans. B, 1990,21, pp. 715–722.

[6] L.H. Shang, F. Paray, J.E.Gruzleski, S. Bergeron, C. Mercadante, C.A. Loong, "Prediction of Microporosity in Al-Si Castings in Low Pressure Permanent Mould Casting Using Criteria Functions," Int. J. Cast Metals Res., 2004,17, pp. 193–200.

[7] S. Fox, and J. Campbell, "Visualisation of Oxide Film Defects During Solidification of Aluminium Alloys," Scripta Mater., 2000, 43, pp. 881–886.

[8] J. Campbell, "An Overview of the Effects of Bifilms on the Structure and Properties of Cast Alloys," Metall. And Mater. Trans. B, 2006, 37, pp. 857–863.

[9] J. Campbell and R.A. Harding, "Casting Technology,"TALAT 2.0 CD-ROM, EAA, Brussels (2000).

[10] J. Campbell, "Castings," Elsevier Science Ltd., Oxford (2003).

[11] E. Fiorese, F. Bonollo, G. Timelli, L. Arnberg, E. Gariboldi, NEW CLASSIFICATION OF DEFECTS AND IMPERFECTIONS FOR ALUMINUM ALLOY CASTINGS, International Journal of Metalcasting, 2015, 9, pp. 55–66.

[12] J. Campbell, "Materials Perspective, Entrainment Defects," Mater. Sci. Technol., 2006, 22, pp. 132–136.

[13] John Campbell, in Complete Casting Handbook (Second Edition), 2015.

[14] Aluminum: Properties and Physical Metallurgy Editor: John E. Hatch, ASM International, ISBN: 978–0–87170-176-3.

[15] K.Gupta, D.J.Lloyd, S.A.Court, Precipitation hardening in Al–Mg–Si alloys with and without excess Si, Materials Science and Engineering: A , 2001, 316, pp.11–17.

[16] H. Liao, Y. Wu, K. Ding, Materials Science and Engineering A, 2013, 560, pp.811–816.

[17] Aluminum Casting Technology, American Foundrymen<s Society, inc., Des Plains, Il, 1986.

[18] Y.F. Ye, Q. Wang, J. Lu, C.T. Liu and Y. Yang, High-entropy alloy: challenges and prospects, Materials Today, 2016, 19, pp. 349–362.

[19] E. J. Pickering and N. G. Jones, High-entropy alloys: a critical assessment of their founding principles and future prospects, International Materials Reviews, 2016, 61, pp. 183–202.

[20] Knuutinen, A., et al., Modification of Al–Si alloys with Ba, Ca, Y and Yb. Journal of Light Metals, 2001. 1(4): p. 229–240.

[21] Nogita, K., et al., Mechanisms of eutectic solidification in Al–Si alloys

modified with Ba, Ca, Y and Yb. Journal of light metals, 2001. 1(4): p. 219–228.

[22] Li, B., F. Kong, and Y. Chen, Effect of yttrium addition on microstructures and room temperature tensile properties of Ti-47 Al alloy. Journal of rare earths, 2006. 24(3): p. 352–356.

[23] Li, H.Z., et al., Effect of Y content on microstructure and mechanical properties of 2519 aluminum alloy. Transactions of Nonferrous Metals Society of China, 2007. 17(6): p. 1194–1198.

[24] Zheng, L. and H. Yongmei, Effect of yttrium on the microstructure of a semi-solid A356 Al alloy. Rare Metals, 2008. 27(5): p. 536–540.

[25] Sheng, M., et al., Effects of Y and Y combined with Al-5Ti-1B on the microstructure and mechanical properties of hypoeutectic Al-Si alloy. JOM, 2015. 67(2): p. 330–335.

[26] Nie, Z.R., et al. Research on rare earth in aluminum. in Materials Science Forum. 2002. Trans Tech Publ.

[27] Nie, Z., et al. Advanced aluminum alloys containing rare-earth erbium. in Materials Forum. 2004.

[28] Xu, G.F., et al., Effect of trace rare earth element Er on Al-Zn-Mg alloy. Transactions of Nonferrous Metals Society of China, 2006. 16(3): p. 598–603.

[29] Li, Y.T., et al. Alloying behavior of rare-earth Er in Al-Cu-Mg-Ag alloy. in Materials science forum. 2007. Trans Tech Publ.

[30] Wen, S.P., et al., The effect of erbium on the microstructure and mechanical properties of Al-Mg-Mn-Zr alloy. Materials Science and Engineering a-Structural Materials Properties Microstructure and Processing, 2009. 516(1–2): p. 42–49.

[31] Colombo, M., E. Gariboldi, and A. Morri, Er addition to Al-Si-Mg-based casting alloy: Effects on microstructure, room and high temperature mechanical properties. Journal of Alloys and Compounds, 2017. 708: p. 1234–1244.

[32] Colombo, M., E. Gariboldi, and A. Morri, Influences of different Zr additions on the microstructure, room and high temperature mechanical properties of an Al-7Si-0.4 Mg alloy modified with 0.25% Er. Materials Science and Engineering: A, 2017.

[33] Li, Q.L., et al. Effect of Rare Earth Er on Microstructure and Mechanical Properties of Cast Al-Si-Mg Alloy. in Materials Science Forum. 2017. Trans Tech Publ.

[34] Xu, C., et al., Effect of Al–P–Ti–TiC–Nd 2 O 3 modifier on the microstructure and mechanical properties of hypereutectic Al–20wt.% Si alloy. Materials Science and Engineering: A, 2007. 452: p. 341–346.

[35] Shi, W.X., et al., Effect of Nd on microstructure and wear resistance of hypereutectic Al-20%Si alloy. Journal of Alloys and Compounds, 2010. 508(2): p. 480–485.

[36] Weixi, S., et al., Effect of neodymium on primary silicon and mechanical properties of hypereutectic Al-15% Si alloy. Journal of Rare Earths, 2010. 28: p. 367–370.

[37] Ren, X., et al., Effect of Nd on microstructure and properties of 2A70 alloy. Journal of Alloys and Compounds, 2017.

[38] Tang, Q., et al., The effects of neodymium addition on the intermetallic microstructure and mechanical properties of Al-7Si-0.3 Mg-0.3 Fe alloys. Journal of Alloys and Compounds, 2018.

[39] Ahmad, R., M. Asmael, and M. Amzar, Effect of ytterbium addition on

microstructure and hardness of Al-6.5Si-1Zn secondary cast alloy. 2006.

[40] Xiao, D., et al., Effect of rare earth Yb addition on mechanical properties of Al–5· 3Cu–0· 8Mg–0· 6 Ag alloy. Materials science and technology, 2007. 23(10): p. 1156–1160.

[41] Zhang, X.M., et al., Effects of Yb addition on microstructures and mechanical properties of 2519A aluminum alloy plate. Transactions of Nonferrous Metals Society of China, 2010. 20(5): p. 727–731.

[42] Li, B., et al., Microstructure evolution and modification mechanism of the ytterbium modified Al-7.5%Si-0.45%Mg alloys. Journal of Alloys and Compounds, 2011. 509(7): p. 3387–3392.

[43] Li, J.H., et al., Refinement of Eutectic Si Phase in Al-5Si Alloys with Yb Additions. Metallurgical and Materials Transactions a-Physical Metallurgy and Materials Science, 2013. 44a(2): p. 669–681.

[44] Hu, Z., et al., Solidification behavior, microstructure and silicon twinning of Al-10Si alloys with Yb addition. Journal of Rare Earths, 2018.

[45] Li, Q., et al., Mechanical Properties and Microstructural Evolution of Yb-Modified Al-20% Si Alloy. Journal of Materials Engineering and Performance, 2018: p. 1–10.

[46] Nogita, K., S.D. McDonald, and A. K. Dahle, Eutectic modification of Al-Si alloys with rare earth metals. Materials Transactions, 2004. 45(2): p. 323–326.

[47] Chen, Z.W., P. Chen, and C.Y. Ma, Microstructures and mechanical properties of Al-Cu-Mn alloy with La and Sm addition. Rare Metals, 2012. 31 (4): p. 332–335.

[48] Qiu, H., H. Yan, and Z. Hu, Effect of samarium (Sm) addition on the microstructures and mechanical properties of Al–7Si–0.7 Mg alloys. Journal of Alloys and Compounds, 2013. 567: p. 77–81.

[49] Zhang, W., et al., Effects of Sc content on the microstructure of As-Cast Al-7wt.% Si alloys. Materials Characterization, 2012. 66: p. 104–110.

[50] Patakham, U., J. Kajornchaiyakul, and C. Limmaneevichitr, Modification mechanism of eutectic silicon in Al–6Si–0.3 Mg alloy with scandium. Journal of Alloys and Compounds, 2013. 575: p. 273–284.

[51] C. Y. Young, L. Qingchaun, and J. Zhuling, "Influence of cerium and mischmetal on the hardness and brightness of Al-Mg-Si alloys," Journal of the Less Common Metals, vol. 110, 1985, pp. 175–178.

[52] M. Ravi, U. T. S. Pillai, B. C. Pai, A. D. Damodaran, and E. S. Dwarakadasa, "A study of the influence of mischmetal additions to ai-7si-0.3 mg (lm 25/356) alloy," Metallurgical and Materials Transactions A, vol. 27, 1996, pp. 1283–1292.

[53] Chang, J.Y., et al., Rare earth concentration in the primary Si crystal in rare earth added Al-21wt.%Si alloy. Scripta Materialia, 1998. 39(3): p. 307–314.

[54] Chang, J.Y., I.G. Moon, and C.S. Choi, Refinement of cast microstructure of hypereutectic Al-Si alloys through the addition of rare earth metals. Journal of Materials Science, 1998. 33(20): p. 5015–5023.

[55] Li, J.H., et al., Modification of eutectic Si in Al-Si alloys with Eu addition. Acta Materialia, 2015. 84: p. 153–163.

[56] Mao, F., et al., The interaction between Eu and P in high purity Al-7Si

alloys. Materials Characterization, 2016. 120: p. 129–142.

[57] Ravi, M., et al., The effect of mischmetal addition on the structure and mechanical properties of a cast Al-7Si-0.3 Mg alloy containing excess iron (up to 0.6 Pct). Metallurgical and Materials Transactions A, 2002. 33(2): p. 391–400.

[58] Weiwei, W.A.N., et al., Study of rare earth element effect on microstructures and mechanical properties of an Al-Cu-Mg-Si cast alloy. Rare Metals, 2006. 25(6): p. 129–132.

[59] Chong, C., et al., Influences of complex modification of P and RE on microstructure and mechanical properties of hypereutectic Al-20Si alloy. Transactions of Nonferrous Metals Society of China, 2007. 17(2): p. 301–306.

[60] El Sebaie, O., et al., The effects of mischmetal, cooling rate and heat treatment on the hardness of A319. 1, A356. 2 and A413. 1 Al–Si casting alloys. Materials Science and Engineering: A, 2008. 486(1): p. 241–252.

[61] El Sebaie , O., et al., The effects of mischmetal, cooling rate and heat treatment on the eutectic Si particle characteristics of A319.1, A356.2 and A413.1 Al–Si casting alloys. Materials Science and Engineering: A, 2008. 480 (1–2): p. 342–355.

[62] Elsebaie, O., F.H. Samuel, and S. Al Kahtani, Intermetallic phases observed in non-modified and Sr modified Al-Si cast alloys containing mischmetal. International Journal of Cast Metals Research, 2013. 26(1): p. 1–15.

[63] Doty, H.W., S.A. Alkahtani, O. Elsebaie and F.H. Samuel, Influence of Metallurgical Parameters on the Impact Toughness of Near Eutectic Al-Si Alloys, in 119th Metalcasting Congress, AFS 2015, Columbus, OH, April 21–23, 2015. 2015, American Foundry Society: Columbus, Ohio.

[64] El Sebaie, O., The Effect of Mischmetal, Cooling Rate and Heat Treatment on the Microstructure and Hardness of 319, 356, and 413 Aluminum-silicon Alloys. 2006, UQAC.

[65] Li, Y.G., et al., Effect of co-addition of RE, Fe and Mn on the microstructure and performance of A390 alloy. Materials Science and Engineering a-Structural Materials Properties Microstructure and Processing, 2009. 527(1–2): p. 146–149.

[66] Zhu, M., et al., Effect of mischmetal modification treatment on the microstructure, tensile properties, and fracture behavior of Al-7.0%Si-0.3%Mg foundry aluminum alloys. Journal of Materials Science, 2011. 46(8): p. 2685–2694.

[67] Zhu, M., et al., Effects of T6 heat treatment on the microstructure, tensile properties, and fracture behavior of the modified A356 alloys. Materials & Design, 2012. 36: p. 243–249.

[68] Mousavi, G.S., M. Emamy, and J. Rassizadehghani, The effect of mischmetal and heat treatment on the microstructure and tensile properties of A357 Al-Si casting alloy. Materials Science and Engineering a-Structural Materials Properties Microstructure and Processing, 2012. 556: p. 573–581.

[69] Mahmoud, M.G., et al., Effect of Solidification Rate and Rare Earth Metal Addition on the Microstructural Characteristics and Porosity Formation in A356 Alloy. Advances in Materials Science and Engineering, 2017: p. 1–15.

[70] Mahmoud, M.G., et al., Effect of Rare Earth Metals, Sr, and Ti Addition on the Microstructural Characterization of A413.1 Alloy. Advances in Materials Science and Engineering, 2017. 2017.

[71] Jiang, W.M., et al., Effects of rare earth elements addition on microstructures, tensile properties and fractography of A357 alloy. Materials Science and Engineering a-Structural Materials Properties Microstructure and Processing, 2014. 597: p. 237–244.

[72] Zhang, J., et al., Microstructural development of Al–15wt.% Mg 2 Si in situ composite with mischmetal addition. Materials Science and Engineering: A, 2000. 281(1): p. 104–112.

[73] Zhang, H.R., et al., Cooling Rate Sensitivity of RE-Containing Grain Refiner and Its Impact on the Microstructure and Mechanical Properties of A356 Alloy. Acta Metallurgica Sinica-English Letters, 2016. 29(5): p. 414–421.

[74] Dang, B., Z.Y. Jian, and J.F. Xu, Effects of rare-earth element addition and heat treatment on the microstructures and mechanical properties of Al-25% Si alloy. International Journal of Materials Research, 2017. 108(4): p. 269–274.

[75] Mahmoud, M.G., A.M. Samuel, H. W. Doty and F.H. Samuel, *Role of Heat Treatment on the Tensile Properties and Fractography of Al–1.2Si–2.4Cu and Al–8.0Si–2.4Cu Cast Alloys Modified with Ce, La and Sr Addition.* International Journal of Metalcasting, 2020. **14**(1): p. 218–242.

[76] Mahmoud, M.G., A.M. Samuel, H. W. Doty and F.H. Samuel, *Effect of the Addition of La and Ce on the Solidification Behavior of Al–Cu and Al–Si–Cu Cast Alloys.* International Journal of Metalcasting, 2020. **14**(1): p. 191–206.

[77] Song, M., K.H. Chen, and L.P. Huang, Effects of Ce and Ti on the microstructures and mechanical properties of an Al-Cu-Mg-Ag alloy. Rare Metals, 2007. 26(1): p. 28–32.

[78] Song, M., D.H. Xiao, and F.Q. Zhang, Effect of Ce on the thermal stability of the Omega phase in an Al-Cu-Mg-Ag alloy. Rare Metals, 2009. 28 (2): p. 156–159.

[79] Song, X.C., H. Yan, and F.H. Chen, Impact of Rare Earth Element La on Microstructure and Hot Crack Resistance of ADC12 Alloy. Journal of Wuhan University of Technology-Materials Science Edition, 2018. 33(1): p. 193–197.

[80] Xiao, D.H., J.N. Wang, and D.Y. Ding, Effect of minor cerium additions on microstructure and mechanical properties of cast Al – Cu – Mg – Ag alloy. Materials Science and Technology, 2004. 20(10): p. 1237–1240.

[81] Voncina, M., et al., Effect of Ce on solidification and mechanical properties of A360 alloy. Journal of Alloys and Compounds, 2011. 509(27): p. 7349–7355.

[82] Voncina, M., et al., Microstructure and grain refining performance of Ce on A380 alloy. Journal of Mining and Metallurgy, Section B: Metallurgy, 2012. 48(2): p. 265–272.

[83] Chen, Z.W., et al., Kinetic nucleation of primary alpha (Al) dendrites in Al-7%Si-Mg cast alloys with Ce and Sr additions. Transactions of Nonferrous Metals Society of China, 2013. 23(12): p. 3561–3567.

[84] Chen, Y., et al., Effects of combinative addition of lanthanum and boron on grain refinement of Al-Si casting alloys. Materials & Design, 2014. 64: p. 423–426.

[85] Ye, L.Y., et al., Influence of Ce addition on impact properties and microstructures of 2519A aluminum alloy. Materials Science and Engineering: A, 2013. 582: p. 84–90.

[86] Yii, S.L. & Norazman, Anas & Nasir, Ramdziah & Anasyida, A.S..

(2016). Microstructural and Mechanical Properties of Al-20%Si Containing Cerium. Procedia Chemistry. 19. 304–310. 10.1016/j.proche.2016.03.015.

[87] Ahmad, R. and M.B.A. Asmael, Influence of Cerium on Microstructure and Solidification of Eutectic Al–Si Piston Alloy. Materials and Manufacturing Processes, 2015. 31(15): p. 1948–1957.

[88] Ahmad, R., et al., Reduction in secondary dendrite arm spacing in cast eutectic Al–Si piston alloys by cerium addition. International Journal of Minerals, Metallurgy, and Materials, 2017. 24(1): p. 91–101.

[89] Mahmoud, M.G., A.M. Samuel, H. W. Doty and F.H. Samuel, *Formation of Rare Earth Intermetallics in Al–Cu Cast Alloys*, in *Light Metals Symposium held at the 149th Annual Meeting and Exhibition, TMS 2020, 23 February 2020 through 27 February 2020*. 2020: San Diego; United States. p. 241–246.

[90] Mahmoud, M.G., E.M. Elgallad, M. F. Ibrahim and F.H. Samuel, *Effect of Rare Earth Metals on Porosity Formation in A356 Alloy.* International Journal of Metalcasting, 2018. **12**(2): p. 251–265.

[91] Ibrahim, M.F., M.H. Abdelaziz, A. M. Samuel, H.W. Doty and F.H. Samuel, *Effect of Rare Earth Metals on the Mechanical Properties and Fractography of Al–Si-Based Alloys.* International Journal of Metalcasting, 2020. **14**(1): p. 108–124.

[92] Ibrahim, A.I., E.M. Elgallad, A.M. Samuel, H.W. Doty and F.H. Samuel, *Effects of heat treatment and testing temperature on the tensile properties of Al–Cu and Al–Cu–Si based alloys.* International Journal of Materials Research, 2018. **109**(4): p. 314–331.

[93] Elgallad, E.M., M.F. Ibrahim, H.W. Doty and F.H. Samuel, *Microstructural characterisation of Al–Si cast alloys containing rare earth additions.* Philosophical Magazine, 2018. **98**(15): p. 1337–1359.

[94] Huang, X. and H. Yan, Effect of trace La addition on the microstructure and mechanical property of as-cast ADC12 Al-Alloy. Journal of Wuhan University of Technology-Materials Science Edition, 2013. 28(1): p. 202–205.

[95] Lu, T., et al., Effects of La addition on the microstructure and tensile properties of Al-Si-Cu-Mg casting alloys. International Journal of Minerals Metallurgy and Materials, 2015. 22(4): p. 405–410.

[96] Qiu, C., et al., Synergistic effect of Sr and La on the microstructure and mechanical properties of A356. 2 alloy. Materials & Design, 2016.

[97] Tang, P., et al., Influence of strontium and lanthanum simultaneous addition on microstructure and mechanical properties of the secondary Al-Si-Cu-Fe alloy. Journal of Rare Earths, 2017. 35(5): p. 485–493.

[98] Ding, J., et al., Effect of Solution Treatment on Microstructure and Mechanical Properties of A356. 2 Aluminum Alloy Treated With Al–Sr–La Master Alloy. Advanced Engineering Materials, 2018.

[99] Li, C., et al., Effects of Ti and La Additions on the Microstructures and Mechanical Properties of B-Refined and Sr-Modified Al–11Si Alloys. Metals and Materials International, 2018: p. 1–10.

[100] Alkahtani, S.A., E.M. Elgallad, M. M. Tash, A.M. Samuel and F.H. Samuel, Effect of rare earth metals on the microstructure ao AL-SI based alloys. Materials, 2017. **9**(1).

[101] Aguirre-De la Torre, E., et al., Mechanical properties of the A356 aluminum alloy modified with La/Ce.

Journal of Rare Earths, 2013. 31(8): p. 811–816.

[102] Wang, S.C., et al. Effects of La and Ce Mixed Rare Earth on Microstructure and Properties of Al-Mg-Si Aluminum Alloy. in Materials Science Forum. 2017. Trans Tech Publ.

[103] Du, J.D., et al., Effect of CeLa addition on the microstructures and mechanical properties of Al-Cu-Mn-Mg-Fe alloy. Materials Characterization, 2017. 123: p. 42–50.

[104] UK Researchers First to Produce High Grade Rare Earths From Coal: @universityofky; 2017 [updated 2017-11-20. Available from: https://uknow.uky.edu/research/uk-researchers-first-produce-high-grade-rare-earths-coal

[105] Mao, F., et al., Effect of Eu addition on the microstructures and mechanical properties of A356 aluminum alloys. Journal of Alloys and Compounds, 2015. 650: p. 896–906.

[106] Tzeng, Y.-C., et al., Effects of scandium addition on iron-bearing phases and tensile properties of Al–7Si–0.6 Mg alloys. Materials Science and Engineering: A, 2014. 593: p. 103–110.

[107] Xu, C., et al., The synergic effects of Sc and Zr on the microstructure and mechanical properties of Al-Si-Mg alloy. Materials & Design, 2015. 88: p. 485–492.

Chapter 5

Recent Advances of High Entropy Alloys: High Entropy Superalloys

Modupeola Dada, Patricia Popoola, Ntombizodwa Mathe, Samson Adeosun, Sisa Pityana, Olufemi Aramide, Nicholus Malatji, Thabo Lengopeng and Afolabi Ayodeji

Abstract

This study reviews the recent technological advancements in manufacturing technique; laser surface modification and material; High Entropy Superalloys. High Entropy Superalloys are current potential alternatives to nickel superalloys for gas turbine applications and these superalloys are presented as the most promising material for gas turbine engine applications.

Keywords: high entropy alloys, high entropy superalloys, nickel superalloys, turbine engine, laser surface modification

1. Introduction

Energy transformation comprises the turbine, which is an inner combustion device and a spinning engine that utilizes water, wind steam, helium and air to produce work [1]. Kaygusuz [2] stated that dams use turbines as an electrical generator producing electricity for residential and industrial consumption. Nonetheless, in 1939, the first jet engine that powered an aircraft was built consisting of the combustion chamber, the turbine and the compressor [3]. This turbine used air as its working fluid in an internal combustion engine and this engine, in turn, removes enough chemical energy to convert it to mechanical energy from the fuel source while using the working fluid to drive the propeller and the engine [4]. Bell and Partridge [5] anticipated that the Joule cycle is a theoretical cycle for gas turbine applications, where both expansion and compression routes take place in a rotating machine [6]. This comprises some reversible processes such as the turbine using the expansion process and fluid friction for an increase in entropy which causes a spontaneous reaction using the compression method in the Brayton cycle [7]. The gas turbine is characterized by extended overhaul intervals, an increased operating speed, less moving parts, availability, low maintenance, reliability, long life span and rugged design [8]. The design of a turbine engine dictates its performance and the performance requirement are determined by the shaft house power developed in certain temperature conditions which may be extreme. Therefore, the need for high-performance materials becomes necessary because one factor which affects the efficiency of the engine; the turbine inlet temperature is made up of materials which are designed to reduce flow losses and must withstand erosion, corrosion and stress at elevated temperatures [9]. According to Reed [10],

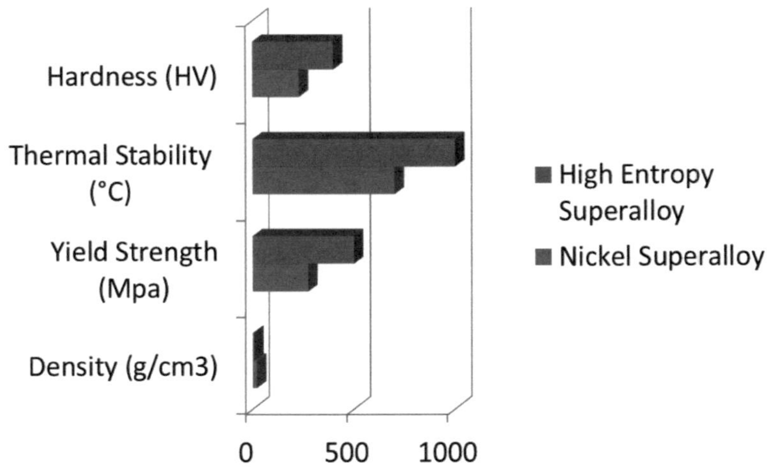

Figure 1.
Comparison graph between nickel super alloy and high entropy superalloy.

superalloys especially Nickel Superalloys are materials generally used at elevated temperatures for these gas turbine applications attributed to their elevated temperature strength, corrosion resistance, excellent formability, cost and low density [11]. However, the nickel-based superalloy has a maximum service temperature, not over 650 $°C$ attributed to the conversion of γ' precipitate strengthening matrix to the δ phase over time [12]. More so, the nucleation and growth of some cavities along the transverse grain boundaries of these materials are the gas turbine airfoil's failure mechanisms [13]. Therefore; a need to develop new materials with improved properties was necessary and this was achieved by transforming conventional material into new ones via advanced industrial reproduction [14]. Miracle, Tsai [15] proposed High Entropy Superalloys (HESAs) as a new class of amalgam with superior properties compared to traditional superalloys as shown in **Figure 1**.

Their elemental composition, lower densities, high configurational entropy and core effects alongside possessing the γ' precipitate reinforcement phase makes this superalloy a preferred alternative material for turbine engine applications [16]. In a previous study, additive manufacturing was presented as a potential advance manufacturing technique as opposed to conventional arc melting and casting fabrication processes. This study attempts to present HESAs as a promising material for gas turbine engine applications, as opposed to traditional Nickel-based superalloys [17].

2. Advances in material development

2.1 Super alloys

Superalloys are stable materials; they do not oxidize or fall apart in very harsh environments and at high temperatures. These amalgams are used for power generation, industrial, marine and aerospace applications [10]. They are characterized by their excellent heat and oxidation resistance at elevated temperatures, high melting temperature, and high-temperature mechanical strength, good fracture toughness, and stress-rupture, creep resistance [18]. In general, superalloys contain more of Co, Ni, Cr or Fe but less of Ta, Hf, W, Cr, B, Mo, Nb, Al, Zr, C, Ti because these elements adversely affect the properties of the blend. Superalloys have a typical face-centred cubic structure and are characterized by a γ' precipitate with operating temperatures above 600 $°C$ [19]. This phase gives the superalloy a principal yield strength which increases with a temperature rise. They may have equiaxed or columnar grain structures without exhibiting high-angle grain boundaries, which at high temperatures are sites for damage accumulation [20]. According to Graybill, Li [21], Superalloy's strengthening mechanism includes dispersion strengthening, solid solution and precipitation strengthening [22]. The dispersion of chemically inert carbides and nitride enhances the strength of the superalloy. Precipitation of all intermetallic phases, namely; carbides and FCC matrix γ' precipitate enhances the strength of the superalloy through Ti, Cb, Ta and Al, which promote the formation of the γ' precipitate. Finally, solid solution strengthening with tungsten, columbium, rhenium, molybdenum, rhenium and tantalum stabilizes the FCC structure and strengthens the superalloy [23].

Gessinger and Bomford [24] suggested that Superalloys for gas turbine applications are widely fabricated using powder metallurgy. However, Bewlay, Gigliotti [25] fabricated the turbine disks using hot die forging and roll forming. Lavella, Berruti [26] studied the residual stresses in Inconel 718 turbine disks fabricated by milling while Groh, Gabb [27] developed a turbine disk using the casting technique. Compared with these conventional techniques, the powder metallurgy process produces turbine disks which are extremely difficult to forge; die life is relatively poor and die fill is extremely difficult but not with additive manufacturing.

2.1.1 Nickel superalloys

Gas turbine engines require higher temperatures for efficiency. This high-temperature application, therefore, requires excellent emission control with an advance in the combustion hardware of the engine. Nickel superalloys materials were developed for this purpose and they make up about half of the weight of materials used in turbine engines [28, 29]. They have an FCC nickel matrix which is stable enough for the alloy to be used for combustion liners, blades, vanes, thermal barrier coatings, burners and are also applied to bear loads of over 75% of their emergent melting temperature. This is attributed to their characteristic high-temperature rupture and creep resistance, lifetime expectancy, low operating costs and excellent thermal efficiency [30]. Nickel superalloys are also used in space vehicles, submarines, petrochemical equipment and nuclear reactors. Nickel-based superalloy 718 (IN718) is widely used in wrought or cast at 540 $°C$ for rotors in gas turbine applications [14]. Nickel superalloy 925 and 725 having good corrosion resistance are applied in the oil and gas industry where carbon dioxide, hydrogen sulphide, free Sulfur and chloride levels are significantly high. Nickel superalloy 706 (IN706) is used for power generation for its large diameter and lower concentrations of other alloying elements. Alloy 685 (Waspaloy) with high-temperature strength and age hardening is widely used for gas turbine engine applications [31]. The superalloy is resistant to corrosion and oxidation whilst withstanding extreme atmospheric conditions while in service. Other compositions are; Rene' N6 used in Jet engines, Inconel Alloy 600 used for stills, condensers, heaters and evaporator tubes. Alloy 601 is used for pollution control, power and aerospace applications [32]. Nimonic 90 is used for turbine disc, blades, hot-working tools and forging. However, Nickel-based superalloy is difficult to machine attributed to its hardness, toughness, and they possess high heat resistance at elevated temperatures alongside low thermal conductivity [33]. Machining at high pressure causes work hardening rapidly, which invariably causes the alloy component to warp. Furthermore, Nickel superalloys can be easily replaced with alloys that have high creep strength and niobium silicide was an appropriate system to replace nickel superalloy having 170 MPa creep strength and at a density of 7 g/cm^3 until recently. Research showed that the addition of Ru and Re to the nickel superalloy led to the enhancement of the superalloy's creep strength, however; these additions are expensive and cause density inversion which results in the defect [34]. This superalloy's stability is limited at very high temperatures [35].

According to Durand-Charre [36], Nickel-based Superalloys are majorly FCC phase structured. However, in aluminum-nickel superalloy systems, a second precipitate phase is formed, which is usually Ni_3Al in a composition containing an ordered intermetallic structure [37]. The γ' phase relies on the cooling rate of the superalloy through the solvus temperature of 894 °C [38]. A fast cooling rate promotes a unimodal distribution of the γ' precipitate, therefore, an increasing the volume of the γ' phase through rapid solidification is essential to the strengthening properties of the superalloy [28]. Although the precipitate morphologies can be modified through heat treatment and other secondary phases observed in Nickel-based superalloys are ordered FCC γ', FCC carbides, ordered body-centred tetragonal γ'' and ordered orthorhombic intermetallic phases [39].

Pollock and Tin [28] did an intensive review on nickel superalloys and the authors stated that the commercial superalloys comprise Co, Cr, W, Mo, Ta, W, Nb, Re, Ti, Al, C, Hf, Y, B and Zr. The yield strength of the nickel superalloy is between 900–1300 MPa at room temperature and the fatigue life at 593 °C is 600 MPa at 10^6 cycles and 10^9 cycles. Additions of Re, Nb, W and Mo can be added for the solid solution strengthening of the superalloy. Y, Ta and Cr contribute to enhance the corrosion and oxidation properties of the superalloy. While Zr, C, Hf and B are carbides or borides forming agents that help enhances the mechanical properties of the superalloy as they are situated at the grain boundaries. The creep rupture life attains about 1100 °C at 137 MPa stress level after 1000 h which is about 90% fraction of the melting point signifying the need for innovative advanced materials having higher melting points for the hottest regions of the turbine engine.

2.1.2 High entropy superalloys (HESAs)

Throughout the years, alloys utilized for commercial reasons were structured by choosing an element which framed the network of the whole component with the addition of essential solutes to the base component [40, 41]. The blends of these combinations were reduced as could reasonably be expected for the immense development of mass intermetallic mixes existing within the molar atomic proportions of these alloys, hence, attaining a 40% mark or more. Along these lines, the intermetallic phases reduce the quality of the alloys while in service [42, 43]. Therefore, a need arose to search for alloys with atomic percentages lesser than 35% and the possibilities of combining many metallic principal elements in several atomic compositions were further investigated [44]. According to Ye, Wang [45], an innovative class of alloys with these attributes was discovered more than a decade ago by mixing multiple principal elements in equimolar or near-equimolar compositions. Yeh, Chen [46] named the alloys 'High Entropy Alloys' (HEAs). The authors defined HEAs as amalgams having compositions with at least five principal metallic elements, with these components having a molar atomic proportion between 5 and 35% [47]. Studies on HEAs have concluded that most HEAs comprise simple FCC, BCC or HCP solid solutions phase attributed to their thigh-entropy effect [48]. Wang, Li [49] suggested that these solid solution phases with little or no intermetallic matrix enable HEAs to have outstanding properties such as strength, extraordinary mechanical and physical properties at cryogenic temperatures, plastic strain, fracture strength and good ductility; they possess elevated-temperature oxidation resistance and excellent work hardenability and have been reported to possess distinctive magnetic and tribological properties [50, 51]. Furthermore, Senkov, Wilks [52] reported that HEAs are exceptional refractory materials and their fatigue-resistance were reported to exceed conventional alloys by Hemphill [53].

In the literature, the development in the solid solution strengthening of High Entropy Alloys (HEAs) and the precipitation hardening properties of the alloys at temperatures above 1100 °*C* , led to the discovery of High Entropy Superalloys (HESAs). Yeh, Tsao [17] stated that these superalloys are simply HEAs with the bulk of γ' precipitates and they are described by their high elongation at room temperature, compressive strength, lower densities, creep resistance and ultimate tensile strengths at elevated temperatures. Tsao, Yeh [54] suggested that High Entropy Superalloy (HESA) is made up of a first elemental content containing at least 35 at.% and each principal reinforcement elemental combination will have a second elemental content of more than 5 at%, for example, $Ni_{40.7}Al_{7.8}Cr_{12.2}Fe_{11.58}Co_{20.6}Ti_{7.2}$ HESA. Senkov, Isheim [55] developed a refractory high entropy superalloy and the authors anticipated that the first and second elemental composition content is derived by the mixing entropy of more than 1.5 R (R is the ideal gas constant) alongside the principal strengthening elemental composition, respectively.

Chen, Chang [56] studied the hierarchical microstructural strengthening of HESAs and the composition of strengthening elements can consist of Cu, Fe, Ti, Zr, Co, V, Al, Nb, Cr and Mn. While the overall structure can comprise Mn, Ni, Fe, Ti, Co, Cr and V while for the grain boundary strengthening; C, B and Hf are added but must not be over 15% of the superalloy's total compositional weight. Refractory elements like Ru, Ta, Re, Mo and W can be added but must also contain less than 15% of the total superalloy's weight [55]. Tsao, Yeh [57] in 2013 recommended the development of superalloys using HEAs microstructure with single phases and an additional second phase for elevated temperatures applications. Yeh, Tsao [17] then fabricated $Ni_{40.7}Al_{7.8}Co_{20.6}Cr_{12.2}$ $Fe_{11.5}Ti_{7.2}$ high entropy superalloy (HESA) via casting method. The authors reported that the microstructure of the composition was stable at elevated temperatures and the superalloy was made up of γ' nanosized precipitates with a density lower than 8 g/cm^3.

Daoud, Manzoni [58] developed $Al_8Co_{17}Cr_{17}Cu_8Fe_{17}Ni_{33}$ HESA using thermos-calc, the authors compared the results with Alloy 800H and IN617. They reported that the HESA had higher tensile strength, this was attributed to two phases; one spherical γ' precipitate which was less than 20 nm after the aging temperature at 700 °*C* and another less than 350 nm with an elongated morphology. He, Wang [59] fabricated $Fe_{94}Co_{94}Ni_{94}Cr_{94}Ti_2Al_4$ HESA with γ' nanosized precipitates to manipulate the thermomechanical properties of HESAs, and they argued that the superalloy had γ' nanosized precipitate with an outstanding yield strength and elongation [60]. According to Xiao, Gregoire [61], scanning alternating current calorimetry can be used to quantify the thermomechanical properties of superalloy.

Wang, Zhou [62] investigated $Al_{0.2}CrFeCoNi_2Cu_{0.2}$ HESA and discovered the γ' nanosized precipitate with 30% elongation. Tsao, Yeh [57] developed seven HESAs using the elements Ni, Fe, Al, Cr, Co, Ti by vacuum arc melting. They stated that the development of the γ precipitates in the superalloy was due to Fe, Cr elements and the γ' matrix, they stated that substituting Ni with Ti enhances the thermal stability of HESAs thus encouraging the γ' matrix and by controlling the elemental compositional partitioning in the middle of the γ-γ' phase, the thermal properties of the high entropy γ matrix can be improved. More so, at elevated temperatures after long term exposures $L1_2$ γ' nanosized precipitates were formed without topological closed packed phases. Gwalani, Soni [63] examined $Al_{0.3}CoCrFeNi_2$ and $Al_{0.3}CoCrFeNi$ HESAs and the authors observed γ' precipitate in the $Al_{0.3}CoCrFeNi$ super alloy until 550 °*C* but were replaced with a B2 phase at 700 °*C* after annealing attributed to the increase in aluminum content [64].

Senkov, Isheim [55] and Li, Lee [65] tested $AlMo_{0.5}NbTa_{0.5}TiZr$ HESA by powder metallurgy and they all observed that the superalloy possessed high thermal stability and yield strength superior to nickel superalloys at 1200 °*C* . Kai, Cheng [66] examined the oxidation behavior of a HESA in O_2 environments. The $Ni_2FeCoCrAl_{0.5}$ HESA oxidation kinetics at 900 °C followed a parabolic-rate law forming scales which was dependent on the oxygen pressure. The results showed that the oxidation rates increased with an increase in oxygen pressure however, the kinetics of mass-loss was observed. Shafiee, Nili-Ahmadabadi [67] designed a wrought HESA using Phacomp and CALPHAD technique. The reports showed that the superalloy comprised of γ' nanosized precipitates with lower densities, excellent workability and high thermal stability than Inconel 718 alloy and Waspaloy. Saito, Chen [68] discussed the influence of heat treatments on HESA microstructural evolution and results showed the cast HESA had coarsened γ' precipitates attributed to microsegregation which decreased the solidus making the γ' solvus unclear. Finally, Zhang, Huo [69] prepared cast $Ni_{48-x}Co_{18}$ $Fe_{9.3}Al_{9.7}Cr_{10.5}Ti_{4.5}Mo_x$ HESA to investigate the mechanical and microstructural properties of the superalloy and they concluded that HESAs exhibits good compressive strength at elevated temperature, elongation and tensile strength at room temperatures than nickel superalloy.

HESAs are stable at elevated temperatures compared with commercial Rene' N6, Udimet 700 and Hastelloy X superalloys, this attributed to their sluggish diffusion and high entropy effect. At high temperatures, Nickel-based superalloys form intermetallic topological closed-packed (TCP) phases rich in Fe-Cr because of the high iron content in less than 100 h at 900 °C and this TCP phases formed is detrimental to the stability of superalloys at high temperatures [70]. However, at 900 °C in more than 200 h, there was no TCP phase observed in $Ni_{40.7}Al_{7.8}Cr_{12.2}$ $Fe_{11.58}Co_{20.6}Ti_{7.2}$ HESA and the γ-γ' microstructure of the superalloy remained stable after isothermal aging for 500 h at 1050 °C [57]. The elevated temperature strength of HESAs has been reported to be higher than that of IN 617. This can be attributed to enhancing the APB energy, increasing the lattice distortion and/or adding refractory Ta and W elements in high concentration to the compositional system.

Yeh and Tsao [71] did a thorough analysis of HESAs with the elemental composition of Fe, C, Al, Mo, Cr, Ti, Ni, Co, Ta, W and Nb. The siderophile element was Nickel while the strengthening element was Nb and C. The authors reported that the HESA's microstructure comprised an FCC, L12 crystal structure and γ' phase while the superalloys had hardness values of 400–470 HV at room temperature. At elevated temperatures, the HESAs hardness values recorded were between 300–350 HV. These values are greater than IN718 under high temperature. The yield strength of HESA at 1000 °C was about 500 MPa. At a strain of 150 MPa under a temperature of about 980 °C , the HESAs showed excellent elevated temperature creep strength when compared with first generational superalloys. The creep strength and fatigue resistance of HESAs is due to the positive lattice misfit of the superalloy [72]. The raft which directionally coarsens the γ' precipitate is corresponding with the stress axis which results in a sluggish motion of dislocation in the γ-γ' precipitate interface, thus hindering the propagation of cracks initiated by fatigue and perpendicular to the same stress axis [73]. The HESAs showed promising thermal stability with a compact protection layer of Cr_2O_3, Al_2O_3 observed on the surface of the HESAs at elevated temperatures, while the densities of the HESAs ranged from 7.78–7.94 g/cm^3 as opposed traditional superalloys that range between 7.8–9.4 g/cm^3 attributed to the high concentration of Cr, Fe, Ti and Al elements [42]. Other superalloys used for turbine engine applications are presented in **Table 1**.

2.2 Protection of superalloys in gas turbine applications

Wee, Do [81] described in a review of the mechanical thermal properties of superalloys and the authors stated that superalloys are required to perform excellently under severe thermal and mechanical stresses. The turbine engine may experience failure attributed to linear and cyclic movements of the pistons, connecting rods, rotors and

SuperAlloy	Composition	Phase Structure	Advantages	Disadvantages	Ref
Iron-Based (Incoloy 800H, Type A-286 alloy, IN903)	IN800H (32Ni-21Cr-1.5Mn-01Si-0.3Ti-0.3Al-01C-bal Fe, wt.%)	γ and γ'	Room-temperature strength, high-temperature strength, Creep, Wear and oxidation resistance	Difficult to machine, poor service performance, susceptible to defects, hot corrosion degradation	[74–76]
Cobalt-Based	(Co-30Ni-11Al-2Ti-5.5 W-2.5Ta-0.1B, at%)	γ, γ' and TCP	High strength at elevated temperatures, corrosion -resistant, thermal shock resistant, easy to machine	Low strength compared to other superalloys,	[77, 78]
Titanium-based (TiAl, Ti6Al4V)	(Ti-48Al-2Cr-2Nb)	γ, FCC $L1_2$	High Strength-toughness and fatigue strength, corrosion-resistant,	Low adhesive, high friction coefficient, low ductility	[79, 80]

Table 1.
Superalloys used for turbine engine applications.

shafts majorly affecting the cascade fluids on the surface of the superalloy [82]. For turbine applications, superalloys comprise elements which are meant for elevated temperature strength required for efficiency [83]. However, these alloying elements may also adversely impact the superalloy's resistance when in this severe environmental conditions over some time. Therefore, there may be a need for additional protection of the superalloy through surface treatments [84]. There are several laser surface modification treatments, namely; laser surface hardening, laser surface heat treatment, Laser alloying, laser shot peening, laser surface dispersing and laser coatings and cladding [85]. Laser coatings enable the superalloy to be resistant to its environment, have microstructural stability and enhance its thermal, physical and mechanical properties [28, 86]. The coatings available can be classified as; overlay coatings, diffusion coatings and ceramic barriers [87, 88]. The deposition of Al from a different external source and diffusing it into the base superalloy to for an external layer is called aluminide or diffusion coating. Bonding an oxidation-resistant alloy which is weak but highly effective on a superalloy to enable surface protection and stability is called overlay cladding, while ceramic barriers are ceramic coatings attached to the surface of a superalloy [89].

3. Advances in manufacturing technology

Technological advancements in surface engineering have replaced conventional methods of surface treatments with laser surface modification (LSM) techniques. The use of lasers in LSM has been reported to produce wear, corrosion, fracture and fatigue resistant HEAs coatings. This is attributed to the energy absorption and rapid solidification of the deposition process, which promotes fine microstructures necessary for surface modification.

According to Wu et al. [90] used laser surface alloying to study the phase evolution and cavitation erosion-corrosion behavior of a HEA coating in distilled water and NaCl solution. The study showed that the alloy's cavitation erosion resistance was enhanced in distilled water but not in NaCl solution due to the corrosion. Zhang et al. [91] fabricated HEA by laser surface alloying to examine the properties of the alloy and they reported that the microhardness property of the coating was thrice the number of the substrate and there were improvements in the wear resistance of the alloy. Huang et al. [92] investigated an equimolar HEA on a titanium alloy substrate using LSM and the results also showed enhancements in the wear resistance of the alloy attributed to the manufacturing route which contributed to the formation of the phases observed in the BCC matrix. Nahmany et al. [93] used an electron beam surface remelting technique to modify two-five component HEAs, and the authors inspected the influence of these surface modification processes on the properties of the alloys. The authors observed a significant increase in the microhardness due to the rapid solidification and cooling process associated with the fabrication technique. From literature, it can be deduced that LSM classified into laser surface remelting, surface amorphisation, laser transformation hardening, shock hardening, laser cladding, laser surface alloying and laser shock peening using different types of lasers can be used to enhance the properties of HEAs [94].

4. Laser surface modification

Laser application in surface modification techniques can be dated back to Albert Einstein who was the first scientist to conceive a stimulated emission in 1917 which today makes lasers applicable [95].

A laser is an abbreviation for "light amplification by stimulated emission of radiation". It is classified into CO_2 and Excimer gaseous lasers, Nd:YAG Solid-state Lasers, Liquid Dye lasers and Yb-doped Fiber. These lasers consist of an optical resonator, a pumping energy outlet and a gain medium. The gain medium is located inside the optical resonator which amplifies a light beam using external energy supplied by the pumping energy outlet. They are classified into dyes, semiconductors or fibers, solid and gaseous states.

Lasers are generally characterized by the ability to avoid divergence in a long-distance, possession of an increased level of energy and monochromaticity [96].

a. The CO_2 laser comprises an electric pump, discharge tube, CO_2 gas for the gain medium and optics such as silver or gold mirrors, zinc selenide lens and finally a window as the optical resonator. Although the Helium-Neon laser was the first gas laser developed in a Bell telephone laboratory, still, the CO_2 is the most widely used gas laser for its high emission wavelength between 9–11 μm which offers very high power for surface modification. The process experiences low light absorption in the infrared regions, reduced optical fiber delivery, instability in the output power attributed to the contraction of the laser structure and thermal expansion when pumping the gas by an AC or DC which sometimes limits its application. Zhang et al. [97] reported fabricating HEAs with CO_2 laser, and the alloy had fine microstructural morphologies and higher mechanical properties. While Zheng et al. [98] mentioned that the HEAs coating fabricated using gas lasers had cellular crystals with dispersion precipitates although the hardness values were reported to be high [99].

b. Excimer lasers, on the other hand, is a mixture of noble gases like helium buffer gas, xenon, argon and a chloride or fluoride halogen. Excimer which is about 248 nm is also known as excited dimers which are pumped using a pulsed electrical discharge for the production of nanosecond pulses in an ultraviolet region, for that reason; it can only be operated in a pulsed mode. Other limitations of this laser are low beam quality, the severity of maintenance and high running cost [100]. Sharma et al. [101] reported using an excimer laser with a wavelength of 248 nm for target ablation during the creation of epitaxial single crystal high entropy ABO_3 perovskite thin films. The authors described how this process was significant in understanding different bonding environments to develop macroscopic responses driven by complex exchange interactions and electron–phonon channels.

c. Nd:Yag which is an acronym for neodymium-doped yttrium aluminum garnet laser is a 1064 nm solid-state laser made up of an active ion and a host from either glass or solid crystalline. It is one of the widely used for the surface modification of HEAs attributed to the ability of its light beam to be transported by flexible optical fibers, consequently increasing its delivery efficiency and compactness [102, 103]. It is also not limited by its mode of transport, which can occur both in pulse and continuous modes. Recently, diode lasers have been substituted for Xenon flash lamps as the pump source to improve the quality of the beam. More so, Nd: YVO_4 is a recent substitute for the Nd:Yag laser due to its wider band absorption, high efficiency and lower operating threshold [104].

d. A Fiber Laser is about 848 nm in wavelength with a rare earth doped fiber used for high power generation due to its increased level of efficiency. The Yb-Doped fiber lasers have excellent electrical-to-optical efficiency with system compactness and high-quality beam. Neodymium, holmium, thulium, dysprosium, erbium and praseodymium are other rare earth elements used as a gain medium in fiber lasers. Fiber lasers are usually pumped with laser diodes; however, they are limited by their light propagation through the optical fiber which greatly influences the guiding medium compared with when the propagation occurs through the air inside the fiber. More so, other factors like the Kerr lens and Raman effects limit the performance of the laser, therefore, optical fibers with polarization maintenance are strongly recommended as the gain medium [105]. Fan et al. [106] examined the influence of fiber laser welding on the mechanical and microstructural proprieties in addition to the solute segregation of a high entropy alloy. The authors reported that the alloy showed dendritic structures with those fabricated using Nd:Yag laser and they observed copper's segregation to the interdendritic region were also attributed to its smaller bonding energies with other elements in the HEA composition, conversely; the alloy showed better hardness and strength compared with the Nd:Yag.

e. Organic liquid dye lasers use organic dyes as the gain medium. These liquid dye with about 50–100 nm compared to solids have a higher density of atoms and they are evenly distributed. These lasers with wide bandwidth are replaceable and are transferred from very intricate regions which are sometimes used as solutes in considerable solvents to develop gain mediums [107]. Coumarin, pyrromethene, exalite, pyridine, styryl and fluorescein are dyes used for pulsed or tunable lasers. Nevertheless, these lasers are limited in applications because they require a large volume of organic solvents for efficiency. Xu et al. [108] used a laser stimulated fluorescence equipment consisting of an organic liquid dye to fabricate a HEA and study the performance of the coatings then the influence of aluminum on the properties of the alloy. The authors stated that the laser technology and the aluminum content enabled the phase transitions, grain refinement and corrosion resistance observed.

f. Other types of lasers are; semiconductor lasers, hybrid laser arc welding and free-electron lasers and the fabrications of HEAs using these lasers are limited in research, hence, should be further explored.

4.1 Laser surface melting (LSM)

This type of surface modification is used for material hardening, electrochemical and tribological resistance and reduction in porosity. An increased rate of heat transfer occurs during the interaction between the substrate and the melted HEAs coating surface, especially during solidification. The rapid solidification and cooling rates invariably produce fine microstructures which also enhances the surface properties of the alloys. Chen et al. [109] used LSM on HEAs and they mentioned that the surface modification process increased the electrochemical and mechanical properties of the alloys. Ochelik et al. [105] found that the solidification rate influences the phases formed using LSM. The fast solidification rates promoted the BCC phase observed which was also responsible for the improved hardness properties of the alloys. Cai et al. [110] also reported observing a BCC solid solution phase and improved microhardness properties after using LSM. The as-remelted HEAs coatings had low wear mass loss showing an improvement in the wear resistance.

4.2 Laser transformation hardening (LTH)

The LTH heats the HEAs coating or films at a very high temperature with an unfocused beam, and then rapid cooling occurs immediately without letting equilibrium phases to form by quenching, as a result, generating very low thermal distortion. This method uses a diode laser or CO_2 to increase the surface properties of the HEAs [111].

4.3 Laser surface alloying (LSA)

This involves the direct injection or pre-placement of additional elements unto the surface of the substrate by a laser source. Rapid solidification occurs with the substrate maintaining its temperature while acting as a heat sink, still the composition of the surface changes [112]. Therefore, re-solidification and rapid quenching follow due to the temperature difference between the surface of the substrate and the treated surface zone. Zhang et al. [113] fabricated HEA coatings by LSA, and the HEA coating had a BCC solid solution phase with improved mechanical and corrosion properties. Jiang et al. [114] fabricated HEAs on a 304 stainless steel substrate and they stated that although the alloy had FCC and BCC phases, the BCC phase was more predominant. The authors also recorded a substantial increase in the hardness with good wear-resistant properties.

4.4 Laser glazing

This method produces a nanocrystalline layer or thin amorphous layer on the surface of the substrate, energy is absorbed into the surface which melts the HEAs coating/films to a certain depth with a laser beam and rapid solidification occurs. This process is achieved using a high power density at a short period enough to create the amorphous structure needed for surface modification [115].

5. Conclusion

High-temperature properties of materials used for turbine engine applications are important for the reduction of fuel consumption, operating costs and pollution. Nickel-based superalloys are widely used due to its strength, resistance to degradation in oxidizing environments, toughness and density. However, Nickel superalloy is not stable at elevated temperatures having a maximum service temperature of 649 °*C* , the superalloy at room temperature has a negative lattice misfit, poor thermal conductivity and difficult to machine. High Entropy Superalloys, with similar γ and γ′ phases as the Nickel-based superalloys, shows high tensile strength than Inconel 617 and Alloy 800 H. The superalloy exhibits good oxidation resistance; have lower densities below 8 g/cm^3, a positive lattice misfit and high yield strength compared to traditional nickel superalloys. Controlling the elemental compositional partitioning between the γ-γ′ in high entropy superalloys makes the thermal stability higher than conventional nickel superalloys and equimolar or near equimolar high entropy alloys. Therefore γ′ precipitate strengthening of solid solution high entropy alloys to form High Entropy Superalloys is currently the most promising material for turbine engine applications. Laser surface modification treatments can be used as a protective mechanism for Superalloys.

Acknowledgements

The authors will like to appreciate the National Laser Center (Laser Enabled Manufacturing Resource Group); Council for Scientific and Research (CSIR) and the Surface Engineering Research Laboratory; Tshwane University of Technology, Pretoria, South Africa for their scientific and technical support during this research.

Author details

Modupeola Dada[1*], Patricia Popoola[2], Ntombizodwa Mathe[3], Samson Adeosun[4], Sisa Pityana[3], Olufemi Aramide[2], Nicholus Malatji[2], Thabo Lengopeng[2] and Afolabi Ayodeji[2]

1 Chemical, Metallurgical and Materials Engineering, Tshwane University of Technology, Pretoria, South Africa

2 Tshwane University of Technology, Pretoria, South Africa

3 Council for Scientific and Industrial Research and Tshwane University of Technology, Pretoria, South Africa

4 University of Lagos, Akoka, Lagos

*Address all correspondence to: dadadupeola@gmail.com

References

[1] Boyce, M.P., *Gas turbine engineering handbook*. 2011: Elsevier.

[2] Kaygusuz, K., *Sustainable development of hydroelectric power.* Energy sources, 2002. **24**(9): p. 803-815.

[3] Langston, L.S., G. Opdyke, and E. Dykewood, *Introduction to gas turbines for non-engineers.* Global Gas Turbine News, 1997. **37**(2): p. 1-9.

[4] Heywood, J.B., *Internal combustion engine fundamentals*. 2018: McGraw-Hill Education.

[5] Bell, M. and T. Partridge, *Thermodynamic design of a reciprocating Joule cycle engine.* Proceedings of the Institution of Mechanical Engineers, Part A: Journal of Power and Energy, 2003. **217**(3): p. 239-246.

[6] Cheng, C.-Y. and C.O.-K. Chen, *Power optimization of an irreversible Brayton heat engine.* Energy sources, 1997. **19**(5): p. 461-474.

[7] Viteri, F. and R.E. Anderson, *Semi-closed brayton cycle gas turbine power systems*. 2003, Google Patents.

[8] Walsh, P.P. and P. Fletcher, *Gas turbine engine*. 2005, Google Patents.

[9] Li, Y. and P. Nilkitsaranont, *Gas turbine performance prognostic for condition-based maintenance.* Applied energy, 2009. **86**(10): p. 2152-2161.

[10] Reed, R.C., *The superalloys: fundamentals and applications*. 2008: Cambridge university press.

[11] Liu, C., et al., *Improved castability of directionally solidified, Ni-based superalloy by the liquid metal cooling process.* Metallurgical and Materials Transactions A, 2012. **43**(2): p. 405-409.

[12] Wang, R.-Z., et al., *Creep-fatigue life prediction and interaction diagram in nickel-based GH4169 superalloy at 650 C based on cycle-by-cycle concept.* International Journal of Fatigue, 2017. **97**: p. 114-123.

[13] Ganji, D.K. and G. Rajyalakshmi, *Influence of Alloying Compositions on the Properties of Nickel-Based Superalloys: A Review*, in *Recent Advances in Mechanical Engineering*. 2020, Springer. p. 537-555.

[14] Akca, E. and A. Gürsel, *A review on superalloys and IN718 nickel-based INCONEL superalloy.* Periodicals of engineering and natural sciences, 2015. **3**(1).

[15] Miracle, D.B., et al., *Refractory high entropy superalloys (RSAs).* Scripta Materialia, 2020. **187**: p. 445-452.

[16] Tsao, T.-K., et al., *High temperature oxidation and corrosion properties of high entropy superalloys.* Entropy, 2016. **18**(2): p. 62.

[17] Yeh, A., et al., *Developing new type of high temperature alloys–high entropy superalloys.* International Journal of Metallurgical & Materials Engineering, 2015. **2015**.

[18] Pint, B.A., K. Unocic, and S. Dryepondt. *Oxidation of superalloys in extreme environments*. in *7th International Symposium on Superalloy*. 2010.

[19] Liu, X., et al., *Effects of Nb and W additions on the microstructures and mechanical properties of novel γ/γ'Co-V-Ti-Based superalloys.* Metals, 2018. **8**(7): p. 563.

[20] Locq, D., et al., *Development of new PM superalloys for high temperature applications.* Intermetallics and superalloys, 2000. **10**: p. 52-57.

[21] Graybill, B., et al. *Additive Manufacturing of nickel-based superalloys*. in *International Manufacturing Science and Engineering Conference*. 2018. American Society of Mechanical Engineers.

[22] Walston, S., et al., *Joint development of a fourth generation single crystal superalloy.* 2004.

[23] Perrut, M., et al., *High temperature materials for aerospace applications: Ni-based superalloys and γ-TiAl alloys.* Comptes Rendus Physique, 2018. **19**(8): p. 657-671.

[24] Gessinger, G.H. and M. Bomford, *Powder metallurgy of superalloys.* International Metallurgical Reviews, 1974. **19**(1): p. 51-76.

[25] Bewlay, B., et al., *Net-shape manufacturing of aircraft engine disks by roll forming and hot die forging.* Journal of Materials Processing Technology, 2003. **135**(2-3): p. 324-329.

[26] Lavella, M., T. Berruti, and E. Bosco, *Residual stress analysis in Inconel 718 milled turbine disk.* International Journal of Machining and Machinability of Materials, 2008. **4**(2-3): p. 181-194.

[27] Groh, J., et al. *Development of a new cast and wrought alloy (René 65) for high temperature disk applications*. in *Proceedings of the 8th International Symposium on Superalloy 718 and Derivatives*. 2014. John Wiley & Sons.

[28] Pollock, T.M. and S. Tin, *Nickel-based superalloys for advanced turbine engines: chemistry, microstructure and properties.* Journal of propulsion and power, 2006. **22**(2): p. 361-374.

[29] Ezugwu, E., J. Bonney, and Y. Yamane, *An overview of the machinability of aeroengine alloys.* Journal of materials processing technology, 2003. **134**(2): p. 233-253.

[30] Reed, R., T. Tao, and N. Warnken, *Alloys-by-design: application to nickel-based single crystal superalloys.* Acta Materialia, 2009. **57**(19): p. 5898-5913.

[31] Kennedy, R., *ALLVAC® 718PLUS™, superalloy for the next forty years.* Superalloys, 2005. **718**(706): p. 1-14.

[32] Devaux, A., et al., *AD730TM-A new nickel-based superalloy for high temperature engine rotative parts.* TMS Superalloys, 2012. **911919**.

[33] Thellaputta, G.R., P.S. Chandra, and C. Rao, *Machinability of nickel based superalloys: a review.* Materials Today: Proceedings, 2017. **4**(2): p. 3712-3721.

[34] Flower, H.M., *High performance materials in aerospace*. 2012: Springer Science & Business Media.

[35] Lehockey, E., G. Palumbo, and P. Lin, *Improving the weldability and service performance of nickel-and iron-based superalloys by grain boundary engineering.* Metallurgical and materials transactions a, 1998. **29**(12): p. 3069-3079.

[36] Durand-Charre, M., *The microstructure of superalloys*. 2017: Routledge.

[37] Sikka, V., et al., *Advances in processing of Ni3Al-based intermetallics and applications.* Intermetallics, 2000. **8**(9-11): p. 1329-1337.

[38] Frank, R.B., C.G. Roberts, and J. Zhang. *Effect of nickel content on delta solvus temperature and mechanical properties of alloy 718*. in *7th international symposium on superalloy*. 2010.

[39] Choudhury, I. and M. El-Baradie, *Machinability of nickel-base super alloys: a general review.* Journal of Materials Processing Technology, 1998. **77**(1-3): p. 278-284.

[40] Gao, M.C. and D.E. Alman, *Searching for next single-phase high-entropy alloy compositions.* Entropy, 2013. **15**(10): p. 4504-4519.

[41] Chatterjee, P., V.M. Athawale, and S. Chakraborty, *Selection of materials using compromise ranking and outranking methods.* Materials & Design, 2009. **30**(10): p. 4043-4053.

[42] Chen, J., et al., *A review on fundamental of high entropy alloys with promising high–temperature properties.* Journal of Alloys and Compounds, 2018. **760**: p. 15-30.

[43] Hummel, R.E., *Understanding materials science: history, properties, applications*. 2004: Springer Science & Business Media.

[44] Cantor, B., et al., *Microstructural development in equiatomic multicomponent alloys.* Materials Science and Engineering: A, 2004. **375**: p. 213-218.

[45] Ye, Y., et al., *High-entropy alloy: challenges and prospects.* Materials Today, 2016. **19**(6): p. 349-362.

[46] Yeh, J.W., et al., *Nanostructured high-entropy alloys with multiple principal elements: novel alloy design concepts and outcomes.* Advanced Engineering Materials, 2004. **6**(5): p. 299-303.

[47] Joseph, J., *Study of direct laser fabricated high entropy alloys*. 2016, Deakin University.

[48] Kuehl, R.O. and R. Kuehl, *Design of experiments: statistical principles of research design and analysis.* 2000.

[49] Wang, Y.P., B.S. Li, and H.Z. Fu, *Solid solution or intermetallics in a high-entropy alloy.* Advanced engineering materials, 2009. **11**(8): p. 641-644.

[50] Gludovatz, B., et al., *A fracture-resistant high-entropy alloy for cryogenic applications.* Science, 2014. **345**(6201): p. 1153-1158.

[51] Mishra, R.K. and R.R. Shahi, *Magnetic characteristics of high entropy alloys.* Magnetism and magnetic materials. IntechOpen, Rijeka, 2018: p. 67-80.

[52] Senkov, O., et al., *Refractory high-entropy alloys.* Intermetallics, 2010. **18**(9): p. 1758-1765.

[53] Hemphill, e.a., *Fatigue behavior of Al0. 5CoCrCuFeNi high entropy alloys.* Acta Materialia, 2012. **60**(16): p. 5723-5734.

[54] Tsao, T.-K., et al., *The high temperature tensile and creep behaviors of high entropy superalloy.* Scientific reports, 2017. **7**(1): p. 1-9.

[55] Senkov, O.N., et al., *Development of a refractory high entropy superalloy.* Entropy, 2016. **18**(3): p. 102.

[56] Chen, Y.-T., et al., *Hierarchical microstructure strengthening in a single crystal high entropy superalloy.* Scientific reports, 2020. **10**(1): p. 1-11.

[57] Tsao, T.-K., A.-C. Yeh, and H. Murakami, *The microstructure stability of precipitation strengthened medium to high entropy superalloys.* Metallurgical and Materials Transactions A, 2017. **48**(5): p. 2435-2442.

[58] Daoud, H., et al., *Microstructure and tensile behavior of Al 8 Co 17 Cr 17 Cu 8 Fe 17 Ni 33 (at.%) high-entropy alloy.* Jom, 2013. **65**(12): p. 1805-1814.

[59] He, J., et al., *A precipitation-hardened high-entropy alloy with outstanding tensile properties.* Acta Materialia, 2016. **102**: p. 187-196.

[60] Zhao, Y., et al., *Thermal stability and coarsening of coherent particles in a precipitation-hardened (NiCoFeCr)*

94Ti2Al4 high-entropy alloy. Acta Materialia, 2018. **147**: p. 184-194.

[61] Xiao, K., et al., *A scanning AC calorimetry technique for the analysis of nano-scale quantities of materials.* Review of Scientific Instruments, 2012. **83**(11): p. 114901.

[62] Wang, Z., et al., *Effect of coherent L12 nanoprecipitates on the tensile behavior of a fcc-based high-entropy alloy.* Materials Science and Engineering: A, 2017. **696**: p. 503-510.

[63] Gwalani, B., et al., *Stability of ordered L12 and B2 precipitates in face centered cubic based high entropy alloys-Al0. 3CoFeCrNi and Al0. 3CuFeCrNi2.* Scripta Materialia, 2016. **123**: p. 130-134.

[64] Borkar, T., et al., *A combinatorial assessment of AlxCrCuFeNi2 (0< x< 1.5) complex concentrated alloys: Microstructure, microhardness, and magnetic properties.* Acta Materialia, 2016. **116**: p. 63-76.

[65] Li, Y., et al., *Microstructure and elevated-temperature mechanical properties of refractory AlMo0. 5NbTa0. 5TiZr High Entropy Alloy fabricated by powder metallurgy.* arXiv preprint arXiv:1801.00263, 2017.

[66] Kai, W., et al., *The oxidation behavior of a Ni2FeCoCrAl0. 5 high-entropy superalloy in O2-containing environments.* Corrosion Science, 2019. **158**: p. 108093.

[67] Shafiee, A., et al., *Development and microstructural characterization of a new wrought high entropy superalloy.* Metals and Materials International, 2019: p. 1-12.

[68] Saito, T., et al., *Effect of Heat Treatments on the Microstructural Evolution of a Single Crystal High-Entropy Superalloy.* Metals, 2020. **10**(12): p. 1600.

[69] Zhang, L., et al., *Microstructure and mechanical properties of precipitation-hardened cast high-entropy superalloys.* Materials Science and Technology, 2020: p. 1-7.

[70] Belan, J., *GCP and TCP phases presented in nickel-base superalloys.* Materials Today: Proceedings, 2016. **3**(4): p. 936-941.

[71] Yeh, A.-c. and T.-K. Tsao, *High-entropy superalloy*. 2019, Google Patents.

[72] Zhang, J., et al., *The effect of lattice misfit on the dislocation motion in superalloys during high-temperature low-stress creep.* Acta Materialia, 2005. **53**(17): p. 4623-4633.

[73] Mughrabi, H., *The importance of sign and magnitude of γ/γ' lattice misfit in superalloys—with special reference to the new γ'-hardened cobalt-base superalloys.* Acta materialia, 2014. **81**: p. 21-29.

[74] Chawla, V., et al., *Corrosion Behavior of Nanostructured TiAlN and AlCrN Hard Coatings on Superfer 800H Superalloy in Simulated Marine Environment.* Journal of Minerals and Materials Characterization and Engineering, 2009. **8**(9): p. 693-700.

[75] Sidhu, T., et al., *Oxidation and hot corrosion resistance of HVOF WC-NiCrFeSiB coating on Ni-and Fe-based superalloys at 800 C.* Journal of Thermal Spray Technology, 2007. **16**(5-6): p. 844-849.

[76] Moody, N., et al., *Temperature effects on hydrogen-induced crack growth susceptibility of iron-based superalloys.* Engineering Fracture Mechanics, 2001. **68**(6): p. 731-750.

[77] Chung, D.-W., et al., *Effects of Cr on the properties of multicomponent cobalt-based superalloys with ultra high γ'volume fraction.* Journal of Alloys and Compounds, 2020: p. 154790.

[78] Makineni, S., B. Nithin, and K. Chattopadhyay, *Synthesis of a new tungsten-free γ–γ' cobalt-based superalloy by tuning alloying additions.* Acta Materialia, 2015. **85**: p. 85-94.

[79] Peters, M., et al., *Titanium alloys for aerospace applications.* Advanced engineering materials, 2003. **5**(6): p. 419-427.

[80] Clemens, H. and W. Smarsly. *Light-weight intermetallic titanium aluminides–status of research and development*. in *Advanced materials research*. 2011. Trans Tech Publ.

[81] Wee, S., et al., *Review on Mechanical Thermal Properties of Superalloys and Thermal Barrier Coating Used in Gas Turbines.* Applied Sciences, 2020. **10**(16): p. 5476.

[82] Goward, G.W., *Current research on the surface protection of superalloys for gas turbine engines.* JOM, 1970. **22**(10): p. 31-39.

[83] Couturier, R. and C. Escaravage, *High temperature alloys for the HTGR Gas Turbine: Required properties and development needs*. 2001.

[84] Sims, C.T., N.S. Stoloff, and W.C. Hagel, *superalloys II*. 1987: Wiley New York.

[85] Zhu, S. and F. Wang, *Nanocrystalline, Enamel and Composite Coatings for Superalloys*, in *Production, Properties, and Applications of High Temperature Coatings*. 2018, IGI Global. p. 160-186.

[86] Levi, C.G., *Emerging materials and processes for thermal barrier systems.* Current Opinion in Solid State and Materials Science, 2004. **8**(1): p. 77-91.

[87] Goward, G., *Progress in coatings for gas turbine airfoils.* Surface and coatings technology, 1998. **108**: p. 73-79.

[88] Galetz, M.C., *Coatings for superalloys*, in *Superalloys*. 2015, InTech. p. 277-298.

[89] Chatterji, D., R. DeVries, and G. Romeo, *Protection of superalloys for turbine application*, in *Advances in corrosion science and technology*. 1976, Springer. p. 1-87.

[90] Wu, C., et al., *Phase evolution and cavitation erosion-corrosion behavior of FeCoCrAlNiTix high entropy alloy coatings on 304 stainless steel by laser surface alloying.* Journal of Alloys and Compounds, 2017. **698**: p. 761-770.

[91] Zhang, S., et al., *Synthesis and characterization of FeCoCrAlCu high-entropy alloy coating by laser surface alloying.* Surface and Coatings Technology, 2015. **262**: p. 64-69.

[92] Huang, C., et al., *Dry sliding wear behavior of laser clad TiVCrAlSi high entropy alloy coatings on Ti–6Al–4V substrate.* Materials & Design, 2012. **41**: p. 338-343.

[93] Nahmany, M., et al., *Al x CrFeCoNi High-Entropy Alloys: Surface Modification by Electron Beam Bead-on-Plate Melting.* Metallography, Microstructure, and Analysis, 2016. **5**(3): p. 229-240.

[94] Tian, Y., et al., *Research progress on laser surface modification of titanium alloys.* Applied Surface Science, 2005. **242**(1-2): p. 177-184.

[95] Herd, R.M., J.S. Dover, and K.A. Arndt, *Basic laser principles.* Dermatologic clinics, 1997. **15**(3): p. 355-372.

[96] Natto, Z.S., et al., *Comparison of the efficacy of different types of lasers for the treatment of peri-implantitis: a systematic review.* International Journal of Oral & Maxillofacial Implants, 2015. **30**(2).

[97] Zhang, H., et al. *Synthesis and characterization of NiCoFeCrAl3 high*

entropy alloy coating by laser cladding. in *Advanced Materials Research*. 2010. Trans Tech Publ.

[98] Zheng, B., Q.B. Liu, and L.Y. Zhang. *Microstructure and properties of MoFeCrTiW high-entropy alloy coating prepared by laser cladding*. in *Advanced Materials Research*. 2013. Trans Tech Publ.

[99] Ye, X., et al., *The property research on high-entropy alloy AlxFeCoNiCuCr coating by laser cladding.* Physics Procedia, 2011. **12**: p. 303-312.

[100] Guo, X., et al., *Corrosion behavior of aluminum in fluoride-containing discharge condition for excimer laser structure application.* Materials Research Express, 2019. **6**(10): p. 106519.

[101] Sharma, Y., et al., *Magnetic anisotropy in single-crystal high-entropy perovskite oxide La (C r 0.2 M n 0.2 F e 0.2 C o 0.2 N i 0.2) O 3 films.* Physical Review Materials, 2020. **4**(1): p. 014404.

[102] Dobbelstein, H., et al., *Direct metal deposition of refractory high entropy alloy MoNbTaW.* Physics Procedia, 2016. **83**: p. 624-633.

[103] Nam, H., et al., *Effect of post weld heat treatment on weldability of high entropy alloy welds.* Science and Technology of Welding and Joining, 2018. **23**(5): p. 420-427.

[104] Rafique, M.M.A., *Additive Manufacturing of Bulk Metallic Glasses and their composites–Recent trends and approaches.*

[105] Ocelík, V., et al., *Additive manufacturing of high-entropy alloys by laser processing.* Jom, 2016. **68**(7): p. 1810-1818.

[106] Fan, Y., et al., *Effect of fiber laser welding on solute segregation and proprieties of CoCrCuFeNi high entropy alloy.* Journal of Laser Applications, 2020. **32**(2): p. 022005.

[107] Rekha, M., N. Mallik, and C. Srivastava, *First report on high entropy alloy nanoparticle decorated graphene.* Scientific reports, 2018. **8**(1): p. 1-10.

[108] Xu, Y., et al., *Microstructure Evolution and Properties of Laser Cladding CoCrFeNiTiAlx High-Entropy Alloy Coatings.* Coatings, 2020. **10**(4): p. 373.

[109] Chen, C., et al., *Influences of laser surface melting on microstructure, mechanical properties and corrosion resistance of dual-phase Cr–Fe–Co–Ni–Al high entropy alloys.* Journal of Alloys and Compounds, 2020. **826**: p. 154100.

[110] Cai, Z., et al., *Microstructure and wear resistance of laser cladded Ni-Cr-Co-Ti-V high-entropy alloy coating after laser remelting processing.* Optics & Laser Technology, 2018. **99**: p. 276-281.

[111] Ion, J., *Laser transformation hardening.* Surface engineering, 2002. **18**(1): p. 14-31.

[112] Manilal, K.M., et al., *A Review on Laser Surface Alloying.* International Research Journal of Engineering and Technology, 2017. **4**(03): p. 4.

[113] Zhang, S., et al., *Laser surface alloying of FeCoCrAlNi high-entropy alloy on 304 stainless steel to enhance corrosion and cavitation erosion resistance.* Optics & Laser Technology, 2016. **84**: p. 23-31.

[114] Jiang, P., et al., *Microstructure and Properties of CeO 2-Modified FeCoCrAlNiTi High-Entropy Alloy Coatings by Laser Surface Alloying.* Journal of Materials Engineering and Performance, 2020: p. 1-10.

[115] Pawlowski, L., *Thick laser coatings: A review.* Journal of thermal spray technology, 1999. **8**(2): p. 279-295.

Section 3

Design of High-Entropy Alloys

Chapter 6

Design and High-Throughput Screening of High Entropy Alloys

Yaqi Wu and Yong Zhang

Abstract

A balanced parameter was proposed to design the high entropy alloys (HEAs), which defined by average melting temperature T_m times entropy of mixing ΔS_m over enthalpy of mixing ΔH_m, $\Omega = T_m\Delta S_m/\Delta H_m$, if Ω is larger than 1.1, we can predict that the entropy is high enough to overcome the enthalpy, and solid solution is likely to form rather than the intermetallic ordered phases. The composition can be further refined by using high-throughput screening by preparing the compositional gradient films. Multiple targets co-sputtering is usually used to prepare the films, and physical masking can separate the samples independently, chemical masking can also applied if possible. One example is the self-sharpening screening by using nanoindentations, the serration behaviors may related to the self-sharpening compositions.

Keywords: designs of high entropy alloys, parameter criterion, high-throughput screening, Multiple targets co-sputtering, chemical masking

1. Introduction

In the process of exploring the formation rules of bulk metallic glasses, Greer put forward the “Confusion Rule”, believing that the more components that make up the alloy system, the more favorable it is to form metallic glass. Then Professor Cantor synthesized more than 20 components of alloys and obtained alloys in solid solution state instead of metallic glasses in order to verify the effectiveness of this principle. On the basis of this development, professor Yeh J. W. from National Tsing Hua University believed that it is the high mixing entropy that stabilizes the solid solution, and thus, the concept of a new kind of alloy, which was called the high-entropy alloy, was put forward. High entropy alloy is an alloy system with five or more equal atomic ratio compositions, providing a new idea for the development of new alloy system, which shows four outstanding effects: high mixing entropy effect, hysteresis diffusion effect, lattice distortion effect and cocktail effect. With the in-depth understanding of HEAs, the non-equimolar complex phases (the matrix is a solid solution) HEAs have been an important part of HEAs. Therefore, the HEA can be divided into the first and the second generations [1]. The first-generation high-entropy alloy is composed of at least five components of equal atomic ratio, and the phase structure is a single-phase solid solution. The second-generation high-entropy alloy is composed of at least four non-equal atomic ratio principal elements, and the phases structure are dual and complex phases. Through composition designing and reasonable preparation and processing technology, high entropy alloys can form a simple structure and show excellent properties, such as high hardness, high strength, high temperature softening resistance, good wear resistance, corrosion resistance and so on, which make the high entropy alloys have a broad application prospect, which has been widely concerned and studied by researchers at home and abroad.

2. Designs of high entropy alloys

When designing alloys, people prefer to get solid solutions. For binary alloys, an element will bring lattice distortion when it enters the lattice of solvent elements as solute atoms, so the form of solid solution can change the properties of the alloy, for example, the strength of the alloy is improved due to the lattice distortion caused by different atomic sizes. The most popular criteria to decide whether two metals are soluble, and form a solid solution, are the classical Hume-Rothery rules. Crystal structure, atomic size difference, electronegativity and valence electron concentration will affect the formation of solid solution. The rules can predict the phase composition of the alloy from the intrinsic characteristic cs. but this is not the case for HEAs: examples exist of concentrated solid solutions of elements having different crystal structures before alloying. For instance, $AlCoCrCu_{0.5}Ni$ forms a single body-centered cubic (BCC) lattice [2].

The formation of alloys is inevitably related to the energy of the system, so when discussing the formation of high-entropy alloys, thermodynamic factors are a basis for the study of high-entropy alloy design. Due to the increase of components, the thermodynamic Gibbs free energy, enthalpy change and entropy change of high-entropy alloys become complicated. In addition to considering the essential issues of alloy formation, the element composition of the alloy is designed to obtain the required properties, such as refractory and high temperature resistance, high strength and toughness, corrosion resistance, and magnetic properties. Various elements have different characteristics, and the interaction between different elements makes the performance of the high-entropy alloy show a compound effect, that is, the "cocktail" effect [3]. Therefore, starting from the essence of the elements, selecting the required properties and forming the alloy system.

The discovery of high-entropy alloys has broken through the limitations of traditional alloy systems, and more alloy systems with better performance have been discovered. At the same time, because the number of elements in high-entropy alloys is greater than that of binary alloys, it also brings a huge workload for alloy design. Generally speaking, High-entropy alloys (HEAs) are broadly defined as alloys containing 5 or more main elements, and the content of each main element is between 5% and 35%. The increase in the types and quantities of constituent elements and the changes in the content of each element combine to form a huge number of alloy systems. Take a multi-component alloy containing 5 main elements as an example, assuming that every 10% content change will produce a different alloy system, then there will be 906 types; take a multi-component alloy containing 3–6 main elements as an example, will produce 592 billion new alloys [4]. How to efficiently design high-entropy alloys and predict phase formation in advance is one of the important factors that promote the development of high-entropy alloys. The factors of phase formation of high-entropy alloys are considered from the perspective of thermodynamics and solid solution formation, and corresponding parameters are proposed to comprehensively reflect phase formation.

2.1 Thermodynamic parameters

2.1.1 Proposal of mixed entropy (ΔS_{mix})

Professor Yeh [5] believes that in a multi-component alloy, as the number of components increases, its mixing entropy (mainly configuration entropy) also gradually increases. The higher mixing entropy reduces the Gibbs free energy of the solid solution phase and promotes the formation of the solid solution phase (especially at higher temperatures), thus forming a high-entropy alloy. The calculation formula of high-entropy alloy mixing entropy is shown in Eq. (1),

$$\Delta S_{mix} = -R\sum_{i=1}^{n} c_i \ln c_i \tag{1}$$

In the formula, R is the molar gas constant, R = 8.314 J/$(mol \cdot K)$; n is the number of components of the multi-component material; c_i is the content of the i-th component (at.%). From the calculation formula of mixing entropy, it can be found that when the atomic ratio of n kinds of components is 1, the mixing entropy value formed by the alloy system is the largest. Therefore, the formula for calculating the mixing entropy of a multi-component alloy with equal atomic ratio is shown in Eq. (2)

$$\Delta S_{mix} = R \ln n \tag{2}$$

Early research believed that $\Delta S_{mix} = 1.5R$ is a necessary condition for the formation of high-entropy alloys, so $\Delta S_{mix} = 1.5R$ is the criterion for dividing high entropy and medium entropy, and $\Delta S_{mix} = 1R$ is the boundary between middle entropy and low entropy. However, subsequent research found that some multi-component alloys have high entropy values, but high-entropy alloys are not formed. Although the configuration entropy of any solid solution will definitely increase with the increase of the elements in the composition, the introduction of additional alloying elements will also increase the possibility of forming a stable intermetallic compound phase. From this point of view, the mixed entropy is a necessary factor affecting the formation of high-entropy alloys, and other parameters are needed to assist in the judgment to comprehensively reflect the law of phase formation.

If kinetic factors are not considered, the choice of phase is controlled by thermodynamics, namely Gibbs free energy. The Gibbs free energy formula is shown in (3)

$$\Delta G_{mix} = \Delta H_{mix} - T\Delta S_{mix} \tag{3}$$

It can be seen from the formula (3) that in addition to the mixing entropy (ΔS_{mix}) has an effect on the phase selection, the mixing enthalpy (ΔH_{mix}) is also one of the factors that need to be considered.

It can be seen from the formula (3) that in addition to the mixing entropy (ΔS_{mix}) has an impact on the phase selection, the mixing enthalpy (ΔH_{mix}) is also one of the factors that need to be considered. Therefore, in 2008, Professor Zhang Yong [6] proposed the ΔH_{mix} parameter, whose formula is shown in (4)

$$\Delta H_{mix} = \sum_{i=1, i \neq j}^{n} c_i c_j \Omega_{ij} \tag{4}$$

Among them, $\Omega_{ij} = 4\Delta H_{mix}^{AB}$, ΔH_{mix}^{AB} is the mixing enthalpy of a two-element (A-B) liquid alloy calculated based on the Miedema model, and is a parameter that represents the chemical compatibility between components.

2.2 Proposal of structural parameters

2.2.1 Proposal of the Delta (δ) parameter

For traditional binary solid solutions, the classic Hume-Rothery criterion can predict the phase composition of alloys from the intrinsic characteristics such as atomic size, crystal structure, valence electron concentration, and electronegativity. However, since the components of high-entropy alloys are mostly 3 to 5, this criterion is no longer applicable. Professor Zhang Yong proposed the Delta (δ) parameter. The Delta (δ) parameter is combined with the mixing entropy (ΔS_{mix}) and the mixing enthalpy (ΔH_{mix}) to jointly predict the phase formation of the multi-component alloy. The formula of Delta (δ) parameter is shown in (5)

$$\delta = 100\sqrt{\sum_{i=1}^{n} c_i\left(1 - \frac{r_i}{\bar{r}}\right)^2} \tag{5}$$

Among them, $\bar{r} = \sum_{i=1}^{n} c_i r_i$, ci and ri are the atomic percentage and atomic radius of the i-th element, respectively. Therefore, the phase formation law of multi-component alloys is proposed. **Figure 1** shows the relationship between ΔH_{mix}-δ-ΔS_{mix} and phase selection.

2.2.2 Proposal of Ω parameter

With the development of more and more new multi-component HEAs, in order to obtain more accurate multi-component HEAs phase formation rules. Entropy and enthalpy cannot independently control the formation of amorphous phase. In the Gibbs thermodynamics equation, only the formation of amorphous is expressed as the synergy and competition of entropy and enthalpy. Considering the factors of mixing entropy, mixing enthalpy and melting temperature, the Ω parameter is proposed [7], and its expression is:

$$\Omega = \frac{T_m \Delta S_{mix}}{\Delta H_{mix}} \tag{6}$$

Among them, ΔS_{mix} is the entropy of mixing, ΔH_{mix} is the enthalpy of mixing, and T_m is the melting point of the mixture. It can be seen from the **Figure 1** that when $\delta \geq 1.1$ and $\Omega \leq 6.6\%$, the multi-component alloy easily forms a solid solution phase.

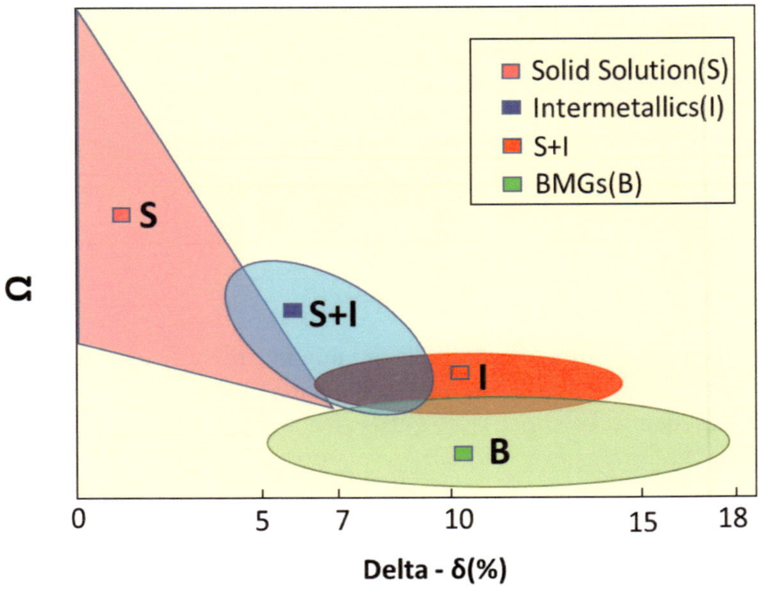

Figure 1.
The relationship between parameters δ and Ω for multi-component alloys [7]. This figure reproducted by the paper "Prediction of high-entropy stabilized solid-solution in multicomponent alloys".

The biggest advantage of this criterion is the combination of size difference, mixing enthalpy and mixing entropy, and the calculation is simple and convenient.

2.2.3 Proposal of electronic structure related parameters (VEC, e/a, ΔX,M_d)

The Hume-Rothery law describes the influence of atomic size, crystal structure, valence electron concentration, and electronegativity on the formation of solid solutions between elements and their laws. Based on the parameters such as δ proposed by this law, the phase formation law of high-entropy alloys was successfully summarized. Can the electronic structure parameters also make a judgment on the solid solution phase?

The related parameters of electronic structure mainly include *VEC*, e/a, ΔX, M_d. It should be emphasized here that both e/a and *VEC* (Valence Electron Concentration) can be called electron concentration. More precisely, e/a is each the average number of electrons flowing in an atom; and *VEC* is the concentration of valence electrons: the total number of valence electrons per atom (including all electrons in the valence band including d electrons).

Guo et al. [8] proposed a relationship between *VEC* and solid-solution stability for multi-component alloys. It should be noted that the *VEC* parameter can only predict phase stability of either FCC or BCC solid-solution in multi-component alloys; But it cannot be used for predicting the formation ability of the solid-solution phase. On the basis of "Ω-δ" rules predicting the formation ability of the solid-solution phase, the *VEC* rule is used to test and verify which type of solid-solution phase is stable in these alloys. When $VEC \geq 8$, it is easy to form solid solution phase of FCC structure; when $VEC < 6.87$, it is easy to form solid solution phase of BCC structure; when $6.87 \leq VEC < 8$, FCC + BCC mixed solid solution phase. *VEC* can also be used to predict the precipitation of σ phase (a TCP phase). The *VEC* criterion is not limited to the determination of the phase stability of as-cast high-entropy alloys. Alloys that have undergone heat treatment or aging treatment can also pass this criterion, but are limited to HEAs containing Cr or V elements [9] (**Figure 2**).

The related parameters of electronic structure can not only predict the law of phase formation, but also guide the design of high-entropy alloy hardness according to the requirements of mechanical properties. When $4.33 \leq VEC \leq 7.55$, HEAs have a single-phase BCC structure, and when $7.80 \leq VEC \leq 9.50$, HEAs have a single-

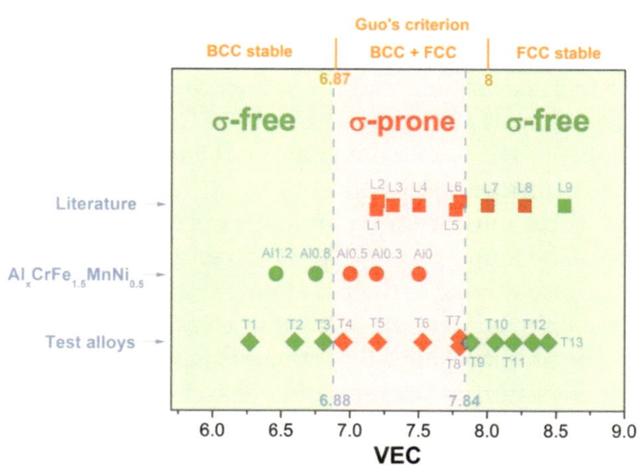

Figure 2.
Relationship between the VEC and the presence of σ phase after aging for a number of HEAs. Green and red icons indicate the absence and presence of σ phase after aging, respectively [9].

phase FCC structure. The study also found that the hardness varies with *VEC* into a Gaussian distribution. When $VEC \approx 6.8$, the hardness is the highest, reaching $650H\nu$. According to the empirical criteria proposed above, HEAs with the required hardness can be designed [10].

In 2014, Wang Zhijun [11] proposed the phase formation criterion of heat-treated HEAs using *VEC* parameters. When $VEC > 7.8$, FCC phase is easy to form, which is less different from the as-cast formation criterion; when $VEC < 6.0$, BCC phase is easy to form, and this area is obviously smaller than as-cast HEAs. Therefore, heat treatment can narrow the range of the BCC solid solution phase.

In addition to the typical body-centered cubic and face-centered cubic phases, there are also TCP phases and Laves phases in high-entropy alloys. These types of phase structures can be predicted by the theories of electronegativity and orbital capacity. Electronegativity ΔX [12] describes the ability of atoms to attract electrons The criterion is not easy to distinguish solid solution phase, intermetallic compound and metallic glass. But it is effectively applied to the formation law of TCP phase in high-entropy alloy. Studies have shown that: when ΔX is small, a solid solution phase is easily obtained, but at the same time intermetallic compounds and metallic glasses can also be formed. Lu Yiping [13] found that: Pauling's electronegativity ΔX has a good correlation with the stability of TCP phase, when $\Delta X > 0.133$, TCP phase is easily formed in HEAs (except those containing a large amount of Al), (Including the σ phase), when $\Delta X < 0.117$, no TCP phase is generated, and when it is between 0.117 and 0.133, the formation of TCP is uncertain.

Different from Pauling's electronegativity above, Polettti [14] proposed electronegativity ΔX_{Allen}, and together with the atomic size difference parameter δ to predict the law of solid solution phase formation, the criterion shows: when δ is between 1% and 6%, and when ΔX_{Allen} is between 3% and 6%, only solid solution phase is formed. Yurchenko [15] used the above two parameters (ΔX_{Allen} and δ) to predict the Laves phase. Through the study of the Laves phase formation law of about 150 kinds of HEAs, it was found that when $\delta > 5\%$ and $\Delta X_{Allen} > 7\%$, there is the Laves phase is generated, and this criterion is not satisfied in only a few cases.

Yurchenko also found that Allen's electronegativity is more accurate than Pauling's electronegativity prediction.

This parameter was first used as a criterion for the formation of TCP phase in Fe-based, Co-based, and Ni-based superalloys. Because HEAs contain a large amount of transition metals, similar to high-temperature alloys, it is used in high-entropy alloys [16] to predict the formation of TCP (including σ phase) phases. (M_d) has a great relationship with the radius and electronegativity of the metal element: it increases with the increase of the radius of the metal element, and decreases with the increase of the electronegativity. As shown in **Figure 3**, this criterion shows that when M_d is greater than 1.09, TCP structure can be generated, but it is not applicable to HEAs containing more Al and V elements, and the related mechanism still needs further study.

In summary, physical and chemical parameters play an important role in phase prediction. It can also be found that configuration entropy is not always the main parameter, there are also mixing enthalpy, atomic size difference, electronic structure, etc. Effective combination of various parameters can improve the accuracy of phase prediction. The physicochemical parameters have certain reference significance, and can roughly estimate the formation phases of multi-component alloys, but because they are derived from experimental data in a certain system, they may not be applicable to other systems, hence a more comprehensive method is needed to overcome system changes. At present, there are more researches on FCC and BCC, but less research on the structure of HCP and orthorhombic crystals.

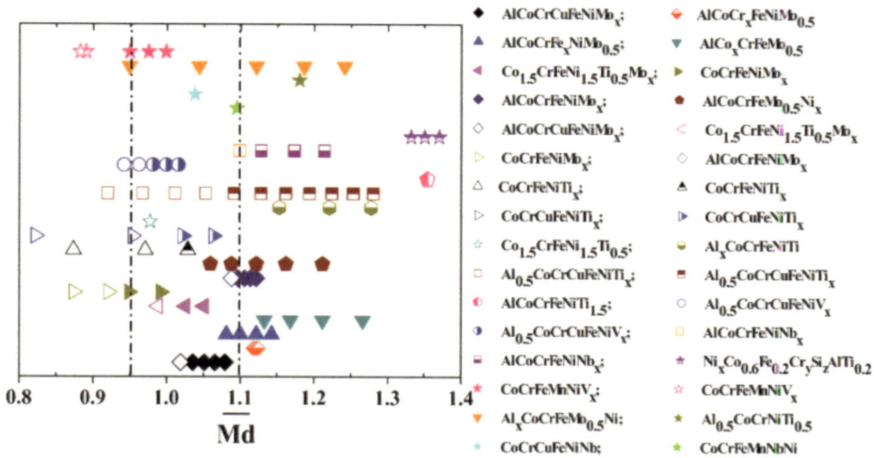

Figure 3.
Relationship between the Md and the TCP phase stability for the HEA systems containing Al, Co, Cr, Cu, Fe, Ni, Mo, Si, Ti, V, Nb, and etc. elements [16].

2.3 Calculation and simulation methods

2.3.1 First principles and molecular dynamics calculations

First-principles refers to the prediction of physical properties without experimental parameters, only calculations by charge, electronic mass and Planck's constant. The first principles mentioned here refer to the first principles based on Density Functional Theory (DFT). This calculation method can systematically study high-entropy alloys from the perspective of theoretical calculations, which is useful for deepening the understanding of HEAs and understanding is very necessary. The exact muffin-tin orbitals (EMTO) based on the framework of density functional theory describes the single-electron potential function and the total energy of the calculation system by using the optimized overlapping potential sphere configuration and the full charge density. This method can not only improve the calculation efficiency, but also ensure sufficient calculation accuracy. The coherent potential approximation (CPA) can solve the problem of the disorder model of multi-principal element replacement solid solution in the first-principles calculation process. Therefore, the combined method of EMTO-CPA is one of the effective methods to solve the first-principles calculation problem of complex alloys with multiple principal elements.

Ab initio molecular dynamics (AIMD) based on quantum mechanics can predict the structure of liquid atoms and better understand the solidification behavior of complex alloys. The AIMD simulation results of Santo Donato et al. [17] showed that during the solidification of the Al1.3CoCrCuFeNi high-entropy alloy, the short-range ordered pairs (Al-Ni, Cr-Fe, and Cu-Cu) in the liquid phase may be B2 ordered phases. At the nucleation point, AIMD can predict the formation trend of short-range ordered phases. Compared with traditional molecular dynamics (MD) simulation, AIMD simulation does not require experiments to obtain the interaction potential between atoms, which is more feasible and more convenient to use. Feng et al. [18] used the DFT calculation method to predict the enthalpy of formation of the actual components in the binary, ternary and quaternary systems. For the $L2_1$ phase, the calculated enthalpy of formation of $AlFe_2Ti$ is -44 kJ/mol, the enthalpy of formation of $AlMn_2Ti$ is -19 kJ/mol, and the enthalpy of formation of $AlCr_2Ti$ is -4 kJ/mol. The $AlFe_2Ti$ phase is a stable ternary phase; When Fe or Mn is used to replace Cr, the corresponding enthalpy of formation increases linearly, and the DFT calculation results are consistent with the experimental results. Ding Xinkai [19] used first-principles methods to predict the phase and structure of NbMoTaWVx high-entropy alloys. The research results show that when $0 \leq x \leq 1.5$, the most stable configuration of $NbMoTaWV_x$ high-entropy alloy in equilibrium is BCC structure; as the component V increases, its density, lattice size and stability of the body-centered cubic phase gradually decrease.

2.3.2 CALPHAD phase diagram calculation

The biggest advantage of the CALPHAD method is to predict and derive the phase diagram of the multi-element system through the binary system or the ternary system composed of it. Unlike traditional thermodynamic databases that only focus on one or two main component regions, HEAs thermodynamic databases need to cover almost the entire component space of a multi-component system. A large number of research results show that the calculation results obtained by the CALPHAD method are basically in agreement with the experimental results, which can be used as a reference for the design of high-entropy alloys [18, 20]. The phase diagram vividly shows the relationship of the various phases in the system. It provides some detailed information of the phase as a function of composition, temperature and pressure. It is a guide diagram for materials science engineering in alloy design and development. The CALPHAD method can be used to study the formation of FCC and BCC phases in multi-component alloys. Since FCC has a greater dynamic effect than BCC structure, the accuracy of using CALPHAD method to predict FCC phase composition is less accurate. The CALPHAD method can also predict the FCC/BCC phase transition of AlxCoCrFeNi high-entropy alloy as cast and heat-treat [21]. Recently, Gao and Senkov used CALPHAD phase diagrams to design a large number of high-entropy alloys. Such as the new HCP structure high-entropy alloys: CoFeMnNi, CuNiPdPt, CuNiPdPtRh [22] and single-phase BCC refractory high-entropy alloys: HfMoNbTiZr, HfMoTaTiZr, NbTaTiVZr, HfMoNbTaTiZr, HfMoTa TiVZr and MoNbTaTiVZr [23]. Therefore, scholars Gao and Senkov believe that CALPHAD is the most direct method for designing HEAs.

2.3.3 Machine learning

Islam [24] uses artificial neural network (ANN) algorithm to predict phase selection, and the accuracy of the neural network model trained by it is 83% on average. As the learning progresses, the accuracy rate gradually increases. The generalization accuracy rates of the four cross-validation data sets were 86.7%, 83.3%, 86.2%, 75.9%, and the average accuracy rate was 83.0%. Although the prediction accuracy of 83% is an acceptable level, a larger data set can infer a higher generalization accuracy. In addition, the neural network parameters after training show that the valence electron concentration *VEC* plays a leading role in determining the phase. Because density functional theory (DFT) calculations are very time-consuming and there are uncertainties in processing the d orbitals of transition metal atoms, HEAs usually contain transition metals. Therefore, Huang [25] uses machine learning algorithms to explore phase formation criteria. It uses the nearest neighbor algorithm (K-nearest neighbors, KNN), support vector machine (Support vector machine, SVM), artificial neural network (Artificial neural network, ANN) 3 algorithms to calculate, respectively, the detection accuracy of 68.6%, 64.3%, 74.3%. When the artificial neural network algorithm is used to distinguish between solid solution + intermetallic compound phase and intermetallic compound phase, the test accuracy reaches 94.3%, which is the best algorithm among the three machine learning algorithms. Huang also evaluated the importance of the five parameters δ, *VEC*, ΔH_{mix}, ΔS_{mix}, $\Delta\chi$ and found that the δ and *VEC* parameters are more important for phase selection.

3. High-throughput screening of high entropy alloys

3.1 Mask method for high-throughput screening

Materials science is an experiment-based science. The traditional experimental methods represented by the "trial and error method" are time-consuming and laborious, and far from meeting the requirements of current industrial development for new materials. With the advancement of technology and commerce more dependent on the advanced materials, the rapid design and development of new materials with excellent performance has become one of the most pressing needs.

The establishment of a systematic study on the design and preparation of high entropy alloys is helpful to the further development of high entropy alloys. The establishment of systematic study on the design and preparation is helpful to the further development of high entropy alloys. For this reason, material genetic engineering provides an effective method to accelerate the screening of the excellent alloy systems of high-performance high entropy alloys, so as to obtain the relationship among material composition, structure and properties with high efficiency, low cost and high reliability.

In addition, compared with traditional alloys, the design, preparation and performance optimization of multi-component materials are more complex, and there is no linear relationship between material properties and mixing entropy, that is, mixed entropy cannot be used as an effective criterion to predict material properties. This makes the screening of multi-component materials more complex and lengthy. The preparation of multi-component gradient materials by high-throughput preparation technology can provide a platform for high-throughput screening of multi-component materials and improve the efficiency of material development.

Different from conventional co-sputtering, the substrate stage was fixed during deposition. A physical mask is placed on the substrate, and the mask can be designed into 2^n areas with the same area according to the requirements. Variations in the relative position between the targets and substrate can cause non-uniformity of deposition. Utilizing the differences in deposition density on the substrate, compositional gradient materials were obtained. The composition of the high entropy alloy film obtained in each small region is different, so its physical and chemical properties are also different. By testing the mechanical properties or corrosion resistance and other chemical properties of the film, the composition range with excellent performance was selected. The high throughput screening method of physical mask provides an efficient and accurate experimental method for the enrichment of high entropy alloy system, and has been widely used in scientific research.

3.2 Serratioins in materials

Serration behavior is a common phenomenon in plastic deformation of materials. For example, serration behavior exists in TRIP steel, amorphous alloys, high-entropy alloys and other materials. The discovery of serration behavior is to explore the relationship between microscopic mechanism and macroscopic properties provides a new way. As the name implies, the stress–strain curve of the material is different from the common type, showing a serration shape. This special mechanical behavior is called the serration of the material. Research shows that serration behavior has a certain similarity with noise, and compares it with noise. This phenomenon occurs when the material undergoes plastic deformation at a certain strain rate, which is caused by dynamic strain aging (DSA). From a microscopic point of view, lattice distortion during plastic deformation leads to a large number of dislocations in the material. The solute atoms interact with the dislocations, and the dislocations are pinned. After the dislocations break free from the pinning, they continue to move. Releasing the microscopic stress caused by lattice distortion has been reported to have similar behavior in low carbon steel.

Temperature, strain rate, load, phase structure, grain size, etc. It is generally believed that it is related to the flow units inside the material. These units can be dislocations, stacking faults, and grain boundaries. The serration behavior is considered to be caused by the interaction with these flow units [26]. Changes in these influencing factors cause the amplitude of the serration to change.

The HEAs possess very high entropy of mixing in their solution states. we cannot distinguish which is the solvent and which is the solute element. In other words, there is no dominant element in HEAs. The special solid solution structure makes defects, in HEAs may be segregated by solute atoms that grants it special plastic-deformation mechanisms. In high-entropy alloys, under high temperature conditions, as the strain rate increases, the sawtooth amplitude of the stress–strain curve becomes smaller and the fluctuation becomes tighter. This may be related to the fact that when the temperature is higher, the atoms are more active, and the thermal motion causes a rapid change between the effect of the atom pinning the defect and the effect of the defect breaking free of the atomic group [27–29]. At lower temperatures and low strain rate, the magnitude of serrations to dramatically increase with an increase in the compression strain. The serrations are believed to be related to the twinning plastic-deformation mechanism [27, 28]. The sample size also has a certain influence on this, Zou et al. have explored it, Such HEA pillars exhibit extraordinarily high yield strengths of about 10 GPa — among the highest reported strengths in micro−/nano-pillar compression and one order of magnitude higher than that of its bulk form — and their ductility is considerably improved (compressive plastic strains over 30%) [30].

3.3 Self-sharpening of materials

Self-sharpening is often used to discuss depleted uranium alloys and grinding wheels. The critical adiabatic shear strain rate and critical adiabatic shear strain value of depleted uranium alloy materials are low, and they are prone to adiabatic shear fracture. They have self-sharpening effects and can penetrate armor. During the armor piercing process, part of the projectile fell off by itself under high-speed collision, exposing a new cutting edge, enhancing the damage. That is a characteristic of the depleted uranium alloy material itself. The self-sharpening of the grinding wheel is the self-sharpening of the structure. During the grinding process, the force acting on the abrasive grains gradually passivates from the sharpness. When the force exceeds the maximum strength of the abrasive grains, the abrasive grains break and form new. When the force exceeds the bonding strength of the bonding agent, the abrasive will fall off the grinding wheel, exposing new abrasive particles, so that it can maintain its own sharpness, which plays an important role in the grinding process.

With the development and progress of modern material technology and the continuous discovery of new materials, many excellent properties of amorphous alloys, such as high strength, corrosion resistance, penetration ability and excellent ferromagnetic properties, are widely used. Amorphous alloy is a new type of alloy, and its internal structure presents a state where chemical disorder and structural disorder coexist. Therefore, the plastic deformation behavior of amorphous alloys has always been a research hotspot. More mature theories believe that different from an alloy with a crystalline structure, which slips through dislocations, however the plastic deformation behavior of amorphous alloys through flow units. The high strength and penetration behavior of amorphous alloys are favored for the preparation of armor piercing materials. For example, the fracture strength of the zirconium-based amorphous alloy is as high as 60, and it has a very high dynamic fracture toughness under the action of high-speed load. It has excellent self-sharpening properties when penetrating the metal, which is better than depleted uranium alloys. And it is a favorable alternative materials [31]. Then the crystallization behavior and size of the amorphous alloy in the manufacturing process are caused, and although the strength of the amorphous alloy is high, the plasticity is poor, so it is restricted.

As we all known, refractory high-entropy alloys (HEAs) posses outstanding mechanical strength at room and high temperature but lack the room temperature ductility. In recent study, Liu et al. [32] designed a new tungsten HEA WFeNiMo with equimolar ratio to enhance chemical disorder, in contrast, to conventional single-principal-element tungsten alloys, and obtained a multi-phase structured alloy. "Self-sharpening", the capacity of a material maintaining its acute head shape during penetration, is a highly required attribute of materials in armor piercing. Depleted uranium kinetic energy penetrators exhibit superior penetration due to their themo-plastic shear banding induced "self-sharpening" behavior during penetration. While tungsten alloy kinetic penetrators usually form mushroom-like heads, which results in a poorer penetration performance when compared to depleted uranium alloy. The development of high-entropy alloys can be said to have opened up new ideas for exploration in the field of materials. In addition to replacing depleted uranium alloys as a projectile, it can even replace drug-type cover materials. According to the theory of penetration fluid dynamics, a long and stable jet can only be formed by a drug-type cover with high density, high sound velocity, favorable thermal conductivity, and high dynamic elongation at break.

The discovery of tungsten-based high entropy alloy breaks through the passivation behavior of traditional tungsten alloy in the process of armor piercing, and maintains the characteristic of high strength, which is expected to replace depleted uranium alloy in the future. However, when selecting the composition system of tungsten-based high-entropy alloy, the test workload and performance testing need to consume a lot of manpower and material resources, such as the traditional Hopkinson rod test or target verification. And what kind of composition ratio will give full play to the performance of tungsten-based high-entropy alloy is an existing problem at present.

So far promising HEAs have been mostly studied in their bulk forms, but small-dimension HEAs have received much less attention. As demands for micro- and nano-scale devices for high-temperature and harsh-environment applications increase, the fabrication and investigation currently popular HEAs at small sizes become more and more interesting. Focused Ga ion beams (FIB) were used by Zou et al. [33] to mill fine-scale pillars out of the obtained HEA films and micro compression tests were carried out using a nano-indenter. After compression a fraction of large pillars, about 1 um in diameter, it was observed that pillars show single slips with wave morphologies or fracture behavior from the SEM images showing in **Figure 4**.

Zou's method provides a new way to solve the problem of "self-sharpening" screening. Combined with the high-throughput screening method of physical mask,

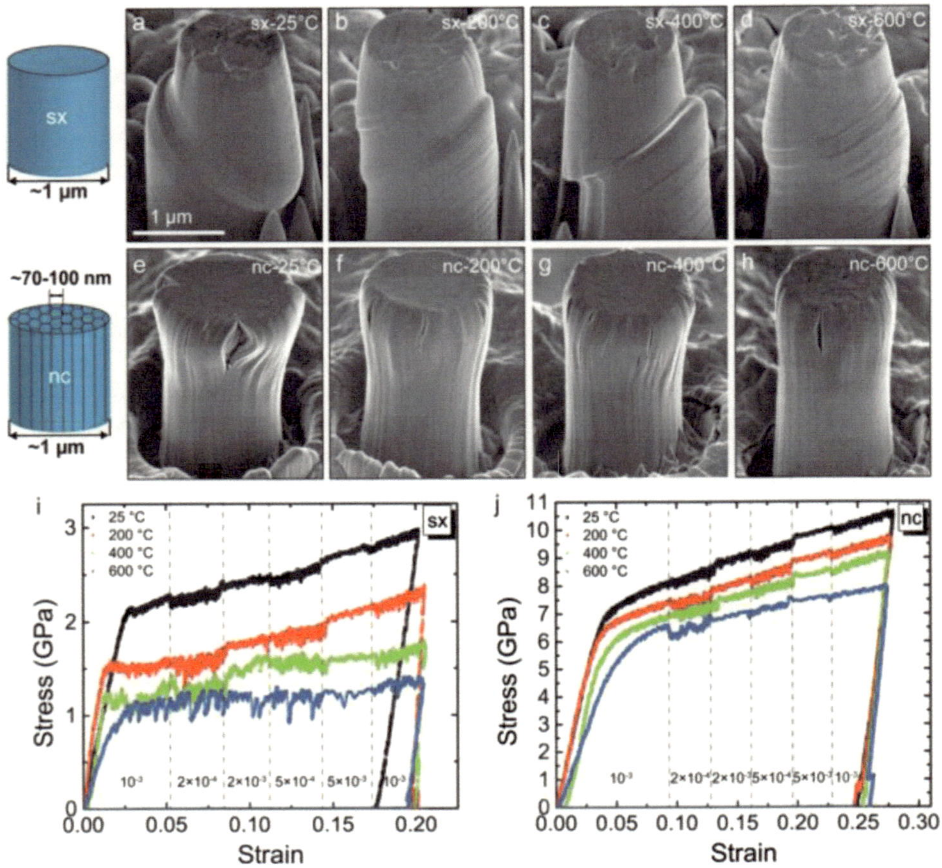

Figure 4.
Compression results for the single-crystalline(sx)- and nanocrystalline(nc)-HEA pillars from room temperature to 600 °C. representative SEM images of the deformed sx-HEA pillars (a – d) and nanostructured columnar-grained HEA pillars (e – h) ([34] copyright@2017 American Chemical Society).

the composition gradient thin films are prepared by magnetron sputtering and high-throughput preparation, and the excellent "self-sharpening" materials can be screened effectively by using Zou's method for reference.

High-throughput screening and nanoscale performance testing can also be used in the design of materials with special needs, such as corrosion-resistant materials. After the corrosion resistance test, the corresponding optimum composition range of the prepared gradient film can be determined, and the film can be prepared in multiple batches and gradually narrow composition range until determined. And combining the above-mentioned zigzag behavior and self-sharpening, there is a certain connection between the two, which may become a hot topic of research in the future.

4. Summary

This section summarizes the parameters for predicting the formation of high-entropy alloy phases and the method of machine learning to predict the formation of high-entropy alloy phases, combined with experimental verification, establishes related databases, and provides data and empirical support for subsequent research on high-entropy alloys. Under continuous research and verification, there are other criteria that continue to emerge, such as: distinguishing single-phase solid solution and multiple solid solution parameters (ϕ) [35], parameters of reaction atom filling mismatch and topological instability (γ), residual stress root mean square $(\varepsilon^2)^{1/2}$ [35, 36], interatomic distance mismatch (S_m) and bulk modulus mismatch (K_m) [37] and other parameters, but these parameters are more practical for some specific alloys, and ΔH_{mix}, δ, Ω, *VEC* et al. parameters are the more mainstream phase criteria, and the range of use is relatively wide. The application of simulation calculation methods such as first principles and machine learning to the research of high-entropy alloys has also achieved remarkable results, providing more accurate and scientific guidance for the composition design of high-entropy alloys. The composition and structure of the material determine the performance of the material. The multi-principal component design makes the phase composition of the high-entropy alloy more complicated. Accurately predicting the formation phase of the alloy with a given composition is very important for the design of high-entropy alloy materials. The current phase prediction requires comprehensive judgment of multiple parameters and multiple methods. The high-throughput screening of high-entropy alloys is also described and the performance of high-entropy alloys in terms of self-sharpening and serrations behavior. Regarding the high-throughput screening method, only the reticle method is described. This method is more practical for the onlooker size. For the screening of the block, the existing research has shown that the method is through hot isostatic pressing. At the same time, matching screening also has high-throughput characterization and high-throughput testing. The description in this chapter is limited. Promoting the development of high-entropy alloys requires more new methods and theories to keep pace with the times.

Author details

Yaqi Wu and Yong Zhang*
State Key Laboratory of Advanced Metals and Materials, University of Science and Technology Beijing, Beijing, China

*Address all correspondence to: drzhangy@ustb.edu.cn

References

[1] Zhang, W., Liaw, P.K. & Zhang, Y. Science and technology in high-entropy alloys. Sci. China Mater. **61,** 2–22 (2018). https://doi.org/10.1007/s40843-017-9195-8

[2] J.W. Yeh, S.J. Lin, T.S. Chin, J.Y. Gan, S.K. Chen, T.T. Shun, C.H. Tsau, S. Y. Chou, Formation of simple crystal structures in Cu–Co–Ni–Cr–Al–Fe–Ti–V alloys with multiprincipal metallic elements, Metall. Mater. Trans. A 35 (8) (2004) 2533–2536. DOI: 10.1007/s11661-006-0234-4

[3] S. Ranganathan, Alloyed pleasures: Multimetallic cocktails, Current Science. 85(2003) 1404e1406. DOI: 10.1038/nature02146

[4] Gorsse S, Couzinie J P, Miracle D B. From high-entropy alloys to complex concentrated alloys. Comptes Rendus Physique, 2018, 19(8): 721–736. DOI: 10.1016/j.crhy.2018.09.004

[5] Yeh J W. Alloy design strategies and future trends in gigh-entropy alloys. JOM, 2013, 65(12):1759–1771. DOI: 10.1007/s11837-013-0761-6

[6] Zhang Y, Zhou Y J, Lin J P, et al. Solid-solution phase formation rules for multi-component alloys. Advanced Engineering Materials, 2008, 10(6): 534–538. DOI:10.1002/adem.200700240

[7] Yang X, Zhang Y. Prediction of high-entropy stabilized solid-solution in multi component alloys. Materials Chemistry and Physics, 2012, 132: 233–238. DOI:10.1016/j.matchemphys.2011.11.021

[8] Guo S, Ng C, Lu J, et al. Effect of valence electron concentration on stability of fcc or bcc phase in high entropy alloys. Journal of Applied Physics, 2011, 109: 103505. DOI: 10.1063/1.3587228

[9] Ming-Hung Tsai, Kun-Yo Tsai, Che-Wei Tsai, Chi Lee, Chien-Chang Juan & Jien-Wei Yeh (2013),Criterion for Sigma Phase Formation in Cr- and V-Containing High-Entropy Alloys, Materials Research Letters, 1:4, 207212, DOI: 10.1080/21663831.2013.831382

[10] Tian F Y, Varga L K, Chen N X, et al. Empirical design of single phase high-entropy alloys with high hardness. Intermetallics, 2015, 58: 1–6. DOI: 10.1016/j.intermet.2014.10.010

[11] Wang Z J, GUO S, Liu C T. Phase selection in high-entropy alloys: from nonequilibrium to equilibrium[J]. JOM, 2014, 66(10): 1966–1972. DOI: 10.1007/s11837-014-0953-8. DOI: 10.1007/s11837-014-0953-8

[12] Guo S, Liu CT. Phase stability in high entropy alloys: formation of solid-solution phase or amorphous phase. Progress in Natural Science: Materials International, 2011, 21: 433–446.

[13] Dong Y, Lu Y P, Jiang L, et al. Effects of electro-negativity on the stability of topologically close-packed phase in high entropy alloys. Intermetallics, 2014, 52: 105–109. DOI: 10.1016/j.intermet.2014.04.001

[14] Poletti M G, Battezzati L. Electronic and thermodynamic criteria for the occurrence of high entropy alloys in metallic systems. Acta materialia, 2014, 75:297–306. DOI: 10.1016/j.actamat.2014.04.033

[15] Yurchenko N, Stepanov N, Salishchev G. Laves-phase formation criterion for high-entropy alloys. Materials Science and Technology, 2016, 33(1): 17–22. DOI:10.1080/02670836.2016.1153277

[16] Lu Y P, Dong Y, Jiang L, et al. A criterion for topological close-packed phase formation in high entropy alloys.

Entropy, 2015, 17: 2355–2366. DOI: 10.3390/e17042355

[17] Santodonato L J, Zhang Y, Feygenson M, et al. Deviation from high-entropy configurations in the atomic distributions of a multi-principal-element alloy[J]. Nature Communications, 2015, 6:5964.

[18] Feng R, Gao M C, Zhang C, et al. Phase stability and transformation in a light-weight high-entropy alloy. Acta Materialia, 2018, 146: 280–293. DOI: 10.1016/j.actamat.2017.12.061.

[19] Ding Kaixi, Sun Kun, Zhang Meng et.al. Research on NbMoTaWVx high-entropy alloy using first-principles calculation method. Journal of Xi'an Jiaotong University, 2018, 52(11):86–92. DOI: 10.7652/xjtuxb201811013

[20] Sheikh S, Mao H H, Guo S. Predicting solid solubility in $CoCrFeNiM_x$ (M= 4d transition metal) high-entropy alloys[J]. Journal of Applied Physics, 2017, 121: 194903. DOI: 10.1063/1.4983762

[21] Zhao XueRou, Lu YuKun, Shi Tuo. Research Progress of High-Entropy Alloy Phase Formation Theory. Materials Reports, 2019, 33(4): 1174–1181. DOI: 10.11896/cldb.17120157

[22] Gao M C, Zhang B, Guo S M, et al. High-entropy alloys in hexagonal close-packed structure. Metallurgical and Materials Transactions A, 2016, 47(7): 3322–3332. DOI: 10.1007/s11661-015-3091-1

[23] Gao M C, Carney C S, Dogan O N, et al. Design of refractory high-entropy alloys. JOM, 2015, 67(11): 2653–2669. DOI: 10.1007/s11837-015-1617-z

[24] Islama N, Huang W J, Zhuang H L. Machine learning for phase selection in multi-principal element alloys. Computational Materials Science, 2018, 150: 230–235. DOI: 10.1016/j.commatsci.2018.04.003

[25] Huang W J, Martin P, Zhuang H L. Machine-learning phase prediction of high-entropy alloys. 2019, 169:225–236. DOI: 10.1016/j.actamat.2019.03.012

[26] Yong Zhang, Jun Peng Liu, Shu Ying Chen, Xie Xie, Peter K. Liaw, Karin A. Dahmen, Jun Wei Qiao, Yan Li Wang. Serration and Noise Behavior in Materials. Progress in Materials Science, 2017, 90. DOI10.1016/j.pmatsci.2017.06.004

[27] Antonaglia J, Xie X, Tang Z, Tsai CW, Qiao JW, Zhang Y, et al. Temperature effects on deformation and serration behavior of high-entropy alloys (HEAs). JOM 2014;66(10):2002–2008.

[28] Laktionova MA, Tabchnikova ED, Tang Z, Liaw PK. Mechanical properties of the high-entropy alloy Ag0.5CoCrCuFeNi at temperatures of 4.2–300 K. Low Temp Phys 2013;39(7):630–632.

[29] Samal S, Rahul MR, Kottada RS, Phanikumar G. Hot deformation behaviour and processing map of Co-Cu-Fe-Ni-Ti eutectic high entropy alloy. Mater Sci Eng, A 2016;664 227–235

[30] Yu Zou, Huan Ma& Ralph Spolenak. Ultrastrong ductile and stable high-entropy alloys at small scales. Na ture Communications,2015, 6(1):7748 DOI:10.1038/ncomms8748

[31] Inoue A, Nishiyama N. New Bulk Metallic Glasses for Applications as Magnetic-Sensing, Chemical, and Structural Materials[J]. MRS Bulletin, 2007, 32(8):651–658.

[32] Xing-Fa Liu, Zhi-Li Tian, Xian-Feng Zhang, Hai-Hua Chen, Tian-Wei Liu, Yan Chen, Yun-Jiang Wang, Lan-Hong Dai, "Self-sharpening" tungsten high-entropy alloy, Acta Materialia,

Volume 186,2020, DOI: 10.1016/j.actamat.2020.01.005

[33] Zou, Yu & Wheeler, Jeffrey & Ma, Huan & Okle, Philipp & Spolenak, Ralph. (2017). Nanocrystalline High-Entropy Alloys: A New Paradigm in High-Temperature Strength and Stability. DOI: DOI: 10.1021/acs.nanolett.6b04716

[34] Ye Y F, Wang Q, Lu J, et al. Design of high entropy alloys: A single-parameter thermodynamic rule. Scripta Materialia, 2015, 104: 53–55.

[35] Wang Z J, Huang Y H, Yang Y, et al. Atomic-size effect and solid solubility of multicomponent alloys. Scripta Materialia, 2015, 94: 28–31.

[36] Ye Y F, Liu C T, Yang Y. A geometric model for intrinsic residual strain and phase stability in high entropy alloys. Acta Materialia, 2015, 94: 152–161.

[37] Caraballo I T, Rivera P E J. A criterion for the formation of high entropy alloys based on lattice distortion. Intermetallics, 2016, 71: 76–87.

Chapter 7

Design and Development of High Entropy Alloys Using Artificial Intelligence

Shailesh Kumar Singh and Vivek K. Singh

Abstract

The conventional design approach of alloys initiates with one principal element and continues by adding several alloying elements to obtain desired properties. In this method, the intrinsic properties of the designed alloy are governed by the principal element. For example, in steel alloy, iron is the principal element, Aluminium in aluminium alloy, and so on. Compared to the conventional alloy, high entropy alloys do not have any dominating elements; all the elements present in these alloys either have an equal or near-equal ratio of elements. As reported in the literature, these alloys exhibit interesting material properties such as high strength, high hardness, improved elevated temperature strength, and magnetic properties. These characteristics make HEAs a suitable option for high-performance applications in the aero engine, aerospace structures, and machine tools. High entropy alloy has multiple principal elements as shown in schematic diagram 1; it leads to much higher possible compositions than conventional alloys. The huge compositional space provides an opportunity to improve desired mechanical properties. If it is explored through "trial and error," it will be challenging and cumbersome. Therefore, search schemes that can competently and promptly recognize particular alloys with desired properties are essential. Artificial Intelligence is a useful tool to model, discover, and optimize new alloys that enable predicting individual material properties as a function of composition. While the application of Artificial Intelligence is quite popular in many aspects of society, its usage in material informatics is still in the nascent stage. The algorithm used in artificial intelligence is trained to pick up predictive rules from data and create a material model quicker than a computational model and can even generate the model for which no physical model exists. Artificial Intelligence (AI) allows predicting a set of experiments to be conducted to detect new alloy having desired properties. Thus, AI can be used as a valuable tool to optimize the development of new alloys.

Keywords: high entropy alloy, Artificial Intelligence, mechanical properties

1. Introduction

Humans have been using alloys since ancient times. Arsenical Bronze, an alloy of copper and arsenic, was used as far back as the 5th millennium BCE. Sumerians first started alloying copper and tin to form Bronze in the third millennium BC. Bronze was much more challenging than its parent elements. Hence, societies wielding bronze weapons skillfully subjugated their neighboring communities. Meteoric iron, a naturally occurring alloy of iron and nickel, was in use from 3200 BCE.

Persians started using carbon steel in the 16th century BCE. With this began the Iron Age. Our modern society also relies heavily on steel.

Typically, alloys have one or two primary components and other components with small amounts. The primary ingredients are chosen for the alloy's leading properties, and the smaller components are selected for specific additional properties. High entropy alloys' development is essential in resolving the limitations of conventional alloys, especially in extreme temperature and loading conditions. High entropy alloys compared to traditional alloys do not have any dominating elements; all the elements present in these alloys either have an equal or near-equal element. As a result, these types of alloy features, as their name suggests, high entropy, i.e., reducing the Gibbs free energy of solid solution and their high configuration entropy mixing is stable at higher temperatures. These characteristics make HEAs a suitable option for high – performance applications such as gas turbines and other aerospace structures, tools, gears, and bearings.

High entropy alloys (HEA) or multi-component alloys are created from equal or similar proportions of multiple elements. There are generally two agreed-upon definitions of HEAs. It should have more than five elements with concentrations ranging from 5–35% of each component, or the entropy of mixing should be greater than 1.5R, where R is the gas constant. Due to each constituent element being a principal constituent, these alloys are characterized by larger configuration entropy. For a 2-component system, the entropy of mixing is given by Eq. (1).

$$\Delta_{mix}\,\mathrm{S} = -nR\left(\mathrm{x_A}\,lnx_\mathrm{A} + \mathrm{x_B}\,lnx_\mathrm{B}\right) \tag{1}$$

From **Figure 1**, the entropy increase is maximum when both the components are of equal proportions. It is easy to follow that for multi-component systems; configurational entropy increases with the increasing number of components. It is maximum for any given number of components when all the constituents' proportions are the same. High entropy alloy can be produced by processing in the gas phase, by methods like sputtering or molecular beam epitaxy, in the liquid phase by methods like arc melting, induction melting, and Bridgman solidification, or in the solid phase by methods like mechanical alloying.

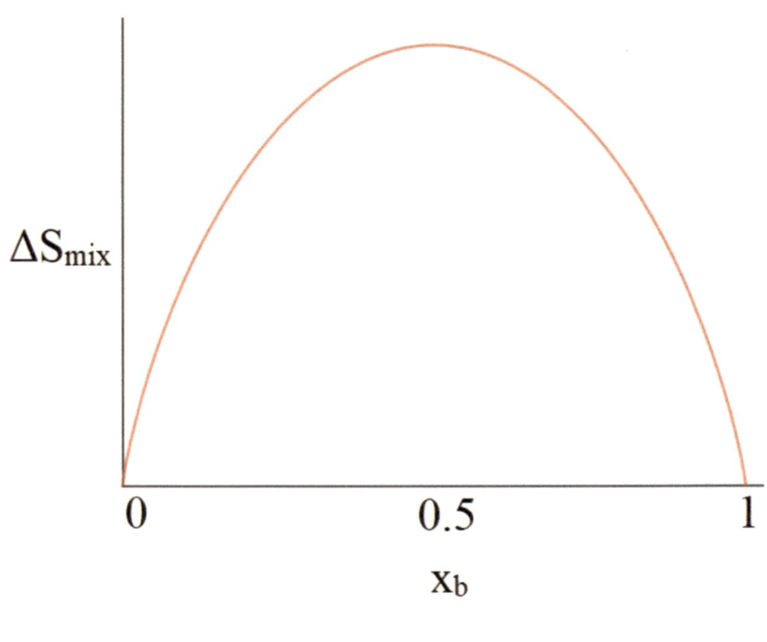

Figure 1.
The illustrated concept of random mixing of elements.

Figure 2 shows tetrahedral compositional space for a 4-element alloy [1]. The corner points represent pure elements. The alloys typically studied lie near the corner points or the edges. One or two primary elements have significant concentration and other elements of minimal concentrations. HEAs are a novel class of materials that lie near the centroid of the compositional space. Research accelerated in this field after the independent publication of Jien-Wei Yeh and Brian Cantor's papers in 2004. HEAs show great strength to weight ratios, fracture resistance, tensile strength, corrosion, and oxidation resistance. HEAs alloy properties are affected by the high-entropy effect, lattice distortion effect, sluggish diffusion effect, and cocktail effect. The crystal structure of HEAs consists of either face-centered cubic (fcc), body-centered cubic (bcc) crystal structure, and hexagonal close-packed (hcp) structure or a mix of any two. VNbMoTaW has excellent refractory properties, CoCrFeMnNi has excellent low-temperature mechanical properties, TaNbHfZrTi shows superconductivity. The reasons for such exceptional properties of HEAs are their microstructures. The most common structures are the Multiphase, single-phase FCC, and single-phase BCC. Generally, when a single-phase HEA is formed, it tends to create a BCC structure, as BCC lattices can accommodate a more extensive range of atomic sizes.

Many techniques are used to explain and guess the different phases of the HEAs based on their parameters, like Valence Electron Concentration (VEC), thermodynamic effects of enthalpy of mixing, and atomic sizes. Phases can be predicted using CALPHAD, molecular dynamics simulations, and kinetic mechanisms. However, using the traditional methods to predict phase formation is very time-consuming and computationally expensive. Density Functional Theory (DFT) is traditionally use to indicate phases by correlating parameters and fitting to existing data. Nevertheless, this method is inadequate given that not much is known about the compositional spaces' central regions. Traditional experimentation with HEAs is done by exploring the compositional and thermodynamic space based on trial-and-error strategies. Recently, Artificial Intelligence (AI) has been used to predict the formation of HEAs. AI use self-learning mechanisms to find patterns in given data. AI is not bound to follow fixed rules in order to obtain the solutions of a specified problem; instead, it solves each unique situation in its way. It is this unique feature

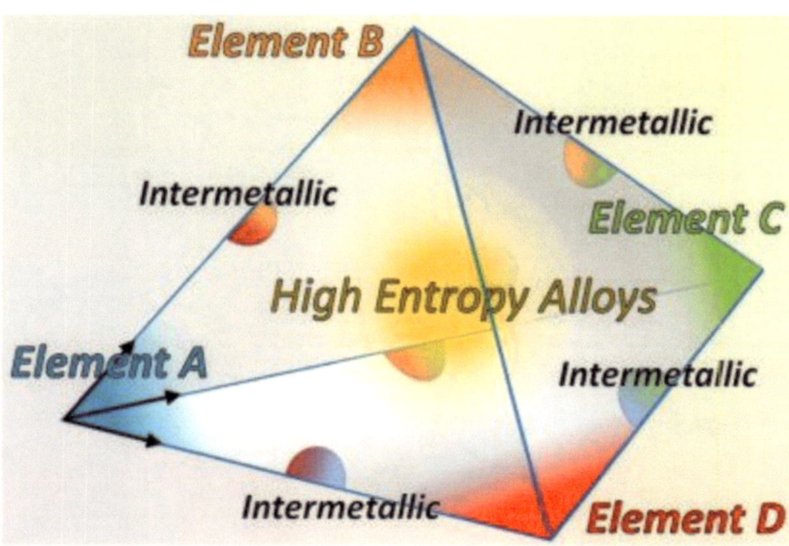

Figure 2.
Compositional space for 4-element alloy [1].

of AI, which can be used to tackle the vagueness of a given system. Phase formation in HEAs is also such a vague system. AI and its subsets, Machine Learning, and Deep Learning can significantly reduce the time component involved in designing HEAs with desired properties by quickly traversing the compositional space.

Adaptive Neuro-Fuzzy Interface System (ANFIS) is an AI algorithm, which is used for phase prediction in FCC, BCC, and multi-phase HEAs. ANFIS is constructed using Artificial Neural Network (ANN) and fuzzy logic. ANN mimics the working of a human brain. Fuzzy logic is a mathematical model that works on degrees of truths, not just absolute truth and false. ANFIS method is suitable to predict the phases in HEAs [2]. Generally two approaches as depicted in **Figures 3** and **4** are used in practice to model the AI problem. The first approach used the constituent elements

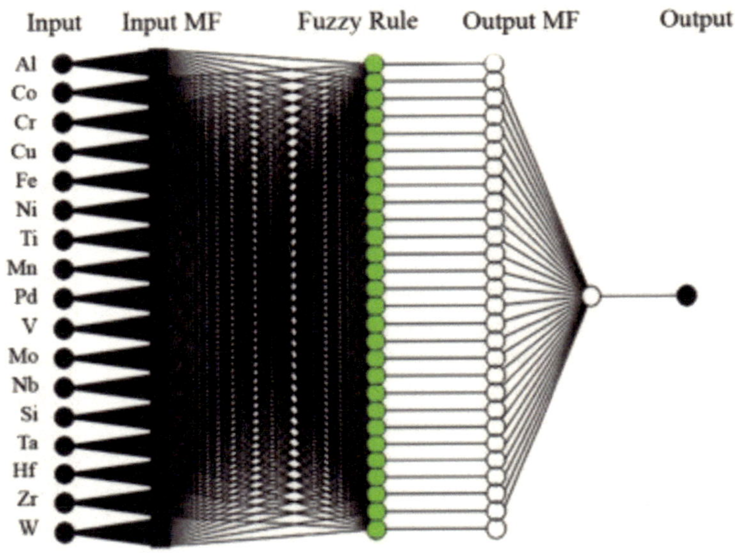

Figure 3.
A framework of ANFIS: Composition-based modeling [2].

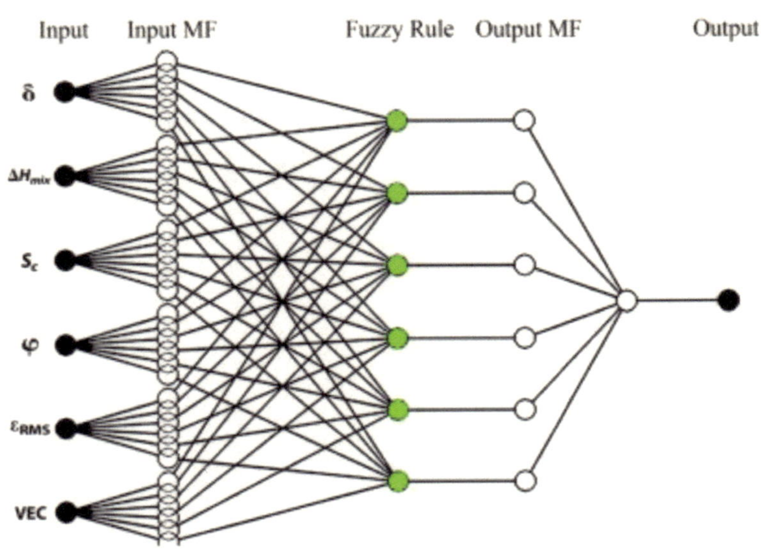

Figure 4.
A framework of ANFIS: Parameter-based modeling [2].

as the inputs, while the second used a set of six crucial parameters in the formation of HEAs. In the first approach, the input elements chosen are Fe, Ni, Ti, Mn, Pd, V, Mo, Nb, Si, Ta, Hf, Zr, and W. In the second approach, atomic size parameter (δ), enthalpy of mixing ($\Delta\Theta_{mix}$), configurational entropy (S_c), single dimensionless thermodynamic parameter (Φ), intrinsic strain (ε_{rms}) and valence electronic concentration (VEC) are the six input parameters. Φ is the thermodynamic parameter; δ and VEC are the parameters of atomic scales; $\Delta\Theta_{mix}$ and S_c are the parameters for nuclear interactions. The first model's accuracy was 84.21%, while it was 80% for the second.

2. Machine learning and design of experiments

Machine Learning (ML) is a subset of Artificial intelligence. ML offers flexibility; as new data becomes available; it is more rapidly able to construct relations between input and output data. Kevin Kaufmann and Kenneth S. Vecchio used an "ML-HEA" to predict the solid solution-forming ability of the HEA using thermodynamic data from ThermoCalc and chemical features with a random forest machine-learning model [3]. They also compared the ML-HEA model with other traditional models like CALPHAD and LTVC, as mention in **Table 1**.

The ML-HEA model correctly predicted the phase for 134 alloy systems with a known phase. This model is in 94% agreement with the CALPHAD model, and 82.1% agreement with the DFT based LTVC model. However, until the central region of the compositional space is further explored, it will be impossible to truly know which of these three models (ML-HEA, CALPHAD, and LTVC) is more accurate in predicting the phase of the HEAs.

Chang et al. used ML with ANN to predict the composition of non-equimolar AlCoCrFeMnNi to get the highest hardness HEAs [4]. In their limited data set of 91 alloys, they also included the alloys of Cu and Mn. The input for the network is the eight elements present in the alloy, and the target output is the hardness of the HEAs. **Figure 5** shows good agreement between the experimental hardness and predicted hardness. Pearson's correlation coefficient = 0.97, R-squared correlation = 0.94, and mean absolute error = 36 Vickers Hardness, and **Table 2** shows good agreement between the predicted hardness and experimental hardness from ML3 to ML5. Here ML3, ML4 and ML5 are high entropy alloy.

Cheng Wen et al. [5] used ML to predict hardness in an AlCoCrCuFeNi HEA system. They used two approaches using the compositions and other descriptive factors such as atomic radii, the difference in electronegativity, VEC, mixing enthalpy, and configurational entropy. They used hardness data of 155 HEAs, with the highest hardness of 735 HV. Pei et al. [6] used machine learning to identify the three most important physical properties of the elements (besides the Hume-Rothery rules) that describe the HEAs' formation. They developed a new method based on these additional features to predict the FCC, BCC or HCP structure of the HEAs.

Material systems	Known solid solution (Neither, FCC, or BCC)	CALPHAD	LVTC
Binaries	117 of 117 (100%)	110 of 117 (94%)	102 of 117 (87.2%)
Ternaries	N/A	362 of 441 (82.1%)	279 of 441 (63.3%)
Quaternaries	8 of 8 (100%)	N/A	690 of 1110 (62.2%)
Quinaries	9 of 9 (100%)	N/A	94 of 130 (72.3%)

Table 1.
Comparison of ML-HEA with CALPHAD and LVTC [3].

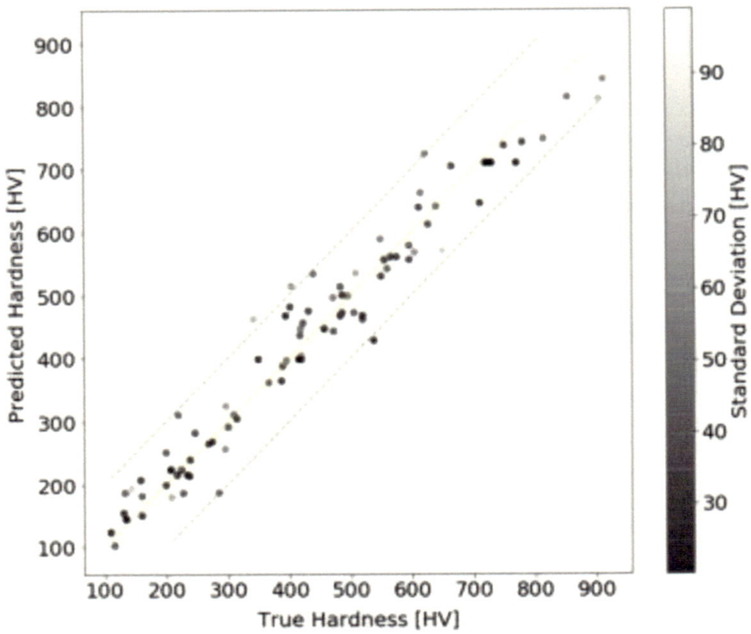

Figure 5.
Predicted versus proper hardness for HEAs, where different colors indicate the variation of the standard deviation [4].

Alloy	Al	Co	Cr	Fe	Mn	Ni	Experimental hardness	Predicted hardness (HV)
Al03	5.66	18.87	18.87	18.87	18.87	18.87	125 ± 3	242 ± 98
ML1	11	18	22	22	5	22	198 ± 6	303 ± 38
ML2	30.5	16	18.5	16.5	5	13.5	522 ± 8	505 ± 35
ML3	30	6	35	6	18	5	605 ± 14	670 ± 94
ML4	25.5	9	35	10	15.5	5	628 ± 13	670 ± 111
ML5	24	18	35	10	7.5	5.5	650 ± 12	670 ± 98

Table 2.
Nominal composition (%) of Alo3 and ML alloys and their hardness values [4].

They used a large dataset of 1252 HEAs to train a model to predict the formation of phases with an accuracy of 93%. For each constituent element of each HEA, they considered 85 elemental properties. From these 85 properties, they can identify the three most essential quantities: molar volume, bulk modulus, and melting temperature. They used these quantities to develop a model based on the physical nature of the elements. This new model was 73% accurate in predicting the phase of the HEAs. **Figure 6** is a hardness map plotted with the variation of the element Al, Co, and Mn contents. The availability of data machine learning technique for alloy design can be a valuable tool to predict HEAs' properties and constituent elements.

Similarly, **Figure 7** used ML and the Design of experiment approach for the accelerated design of HEAs. The first machine-learning model is trained with an existing dataset consisting of hardness and composition of constituent elements. The model is applied to a search in the materials space for unknown value hardness. Based on hardness predictions, a function to select the next set of experiments is performed—this helps to choose an alloy for experimentation.

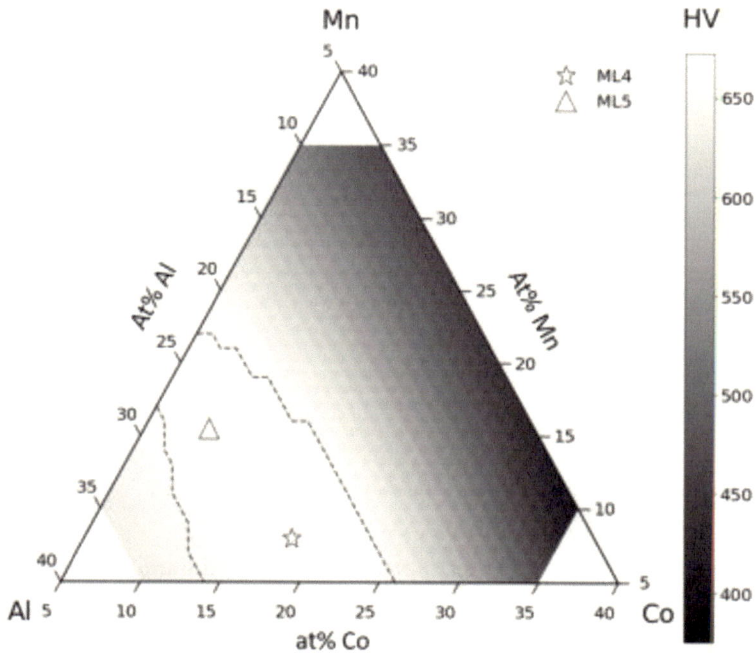

Figure 6.
Hardness plot with a variation of Al, Co, and Mn contents and fixed Co content (black dashed line is the boundary of<30% risk) [6].

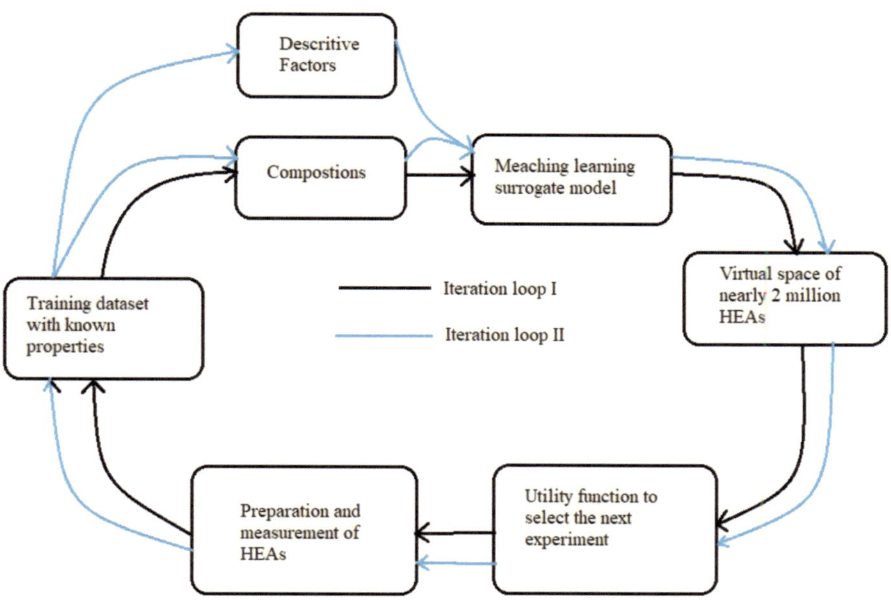

Figure 7.
A schematic of the ML and DOE based iterative design loop for accelerated design of HEAs.

3. Methodology for implementation of AI in HEAs

Figure 8 described the proposed methodology for the implementation of AI to develop a new high entropy alloy. AI algorithm create an AI model by training on the experimental dataset for training, and then this trained AI model is fed with the

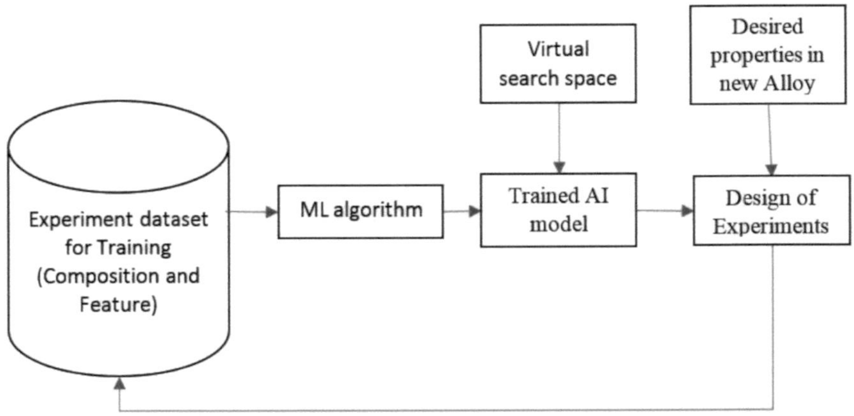

Figure 8.
Methodology for the implementation of AI for the discovery of new alloy.

desired properties of the new model. The output of the AI model is a set of experiments for the creation of a new alloy. The set of experiments suggested by AI model is a subset of an extensive set that will result from all possible combinations of input parameters. Presently trial and error method is used to narrow down this search space, but the use of AI model can lead to selecting a more appropriate subset from this search space. Further, to evolve the AI model with time output of each suggested experiment should be added to the experimental dataset for training and a new AI model is created after each update of the dataset.

The strategy to design and develop of high entropy alloy for achieving the desired strength in the material using the artificial intelligence approach is described below with a detailed explanation:

a. Dataset creation: Data of experiments conducted to create new HEA can be obtained from available literature to create a dataset. This database need to be divided into three subsets (60:20:20): training, validation, and test dataset.

b. Selection of machine learning (ML) algorithm: For ML algorithm, various options like linear regression model, support vector machine, artificial neural network, and deep neural networks should be explored. Starting with the most straightforward algorithm viz. linear regression, each algorithm's performance after training should be evaluated on training and validation dataset (via a loss function like RMSE for expected and actual value of hardness) using k-fold cross-validation. Based on its performance and complexity the ML algorithm can be selected.

c. Training of AI model: AI model created based on selecting the ML algorithm is to be fed with known composition data from the training dataset. The output of this model is the estimated value for the hardness of HEA. Several epochs of training is to be needed to arrive at an acceptable trained AI model.

d. Apart from composition data several physical properties of constituent elements like the difference in atomic radii between composition elements, the difference in electronegativity between components, the valence electron concentration, the mixing enthalpy, the configurational entropy, the U parameter (which is related to the entropy, enthalpy, and the melting point), the L parameter (which is associated with an atom's configuration on a lattice and its radius) and the g parameter (the solid angles of atomic packing for the elements with the most significant and most minor atomic sizes) are related to the intrinsic properties and affect the final hardness. So further AI model should also be trained by accepting both composition data and physical properties as inputs.

e. Training and test datasets with bootstrapping should be used for training the AI model. A bootstrap training set can be generated by resampling the original training data and testing data with replacement. On each made training set, an AI model is created. Each of these AI models can then run on the original training dataset to give different hardness values for each training sample. This give the output as estimated hardness and standard deviation for the estimated hardness.

f. Use of AI model in the design of experiments: By assuming an available range of values for each constituent element, a virtual search space can be created for experiments to conducted of creation of new HEA. Using this virtual space as input to generated AI models, an estimate for hardness and corresponding confidence (standard deviation) can be computed. Candidates in virtual space can then be selected for desired hardness and sorted standard deviation value for the design of experiments.

4. Summary

Artificial Intelligence is a valuable tool to model, discover, and optimize new alloys that enable predicting individual material properties as a function of composition. Artificial Intelligence (AI) allows the prediction of a set of experiments to be conducted to discover new alloy having desired properties. Thus, AI can be used as an effective tool to optimize the development of new alloys. The use of AI for predicting phase, hardness, and other HEAs properties is a gateway to many possibilities. It is the flexibility of AI to adapt to new data that enables further improvement of the models' accuracy, which will lead to a better understanding of phases of HEAs.

Synthesizing HEAs is a complicated and expensive process, so using traditional modeling to explore compositional space and synthesizing each alloy is impractical. AI can be used so that it is not necessary to explore the entire composition space. AI models can handle voluminous experiment data and are bound to outperform human counterparts. Therefore, AI can be effectively employed to competently and promptly recognize particular High Entropy alloys with desired mechanical properties.

Author details

Shailesh Kumar Singh[1*] and Vivek K. Singh[2]

1 Advanced Tribology Research Centre, CSIR-Indian Institute of Petroleum, Dehradun, India

2 Department of Mechanical Engineering, Indian Institute of Technology, Bombay, India

*Address all correspondence to: sk.singh@iip.res.in

References

[1] E. P George, D. Raabe, and R O Ritchie: High Entropy alloy, Nature Review, Vol.4, (2019), pp, 515-534

[2] A.Agarwal, and A. K. Prasada Rao: Artificial Intelligence Predicts Body-Centered-Cubic and Face-Centered-Cubic Phases in High-Entropy Alloys. JOM , Vol. 71, (2019) pp.3424-3432.

[3] K Kaufmann, and K. S. Vecchio: Searching for high entropy alloys: A machine learning approach, Acta Materialia, Volume 198, 2020, pp. 178-222.

[4] Y J Chang, C Y Jui, W J Lee: Prediction of the Composition and Hardness of High-Entropy Alloys by Machine Learning. JOM, Vol. 71, 2019, pp.3433-3442

[5] Cheng Wen, Yan Zhang, Changxin Wang, Dezhen Xue, Yang Bai, Stoichko Antonov, Lanhong Dai, Turab Lookman, and Yanjing Su: Machine learning assisted design of high entropy alloys with desired property, Acta Materialia, Volume 170, 2019, pp. 109-117.

[6] Pei, Z., Yin, J., Hawk, J.A. et al. Machine-learning informed prediction of high-entropy solid solution formation: Beyond the Hume-Rothery rules. npj Comput Mater 6, Vol.50 2020, pp.1-8.